Warfare and Society in the Ancient Eastern Mediterranean

Papers arising from a colloquium held at the
University of Liverpool, 13th June 2008

Edited by

Stephen O'Brien
Daniel Boatright

BAR International Series 2583
2013

Published in 2016 by
BAR Publishing, Oxford

BAR International Series 2583

Warfare and Society in the Ancient Eastern Mediterranean

ISBN 978 1 4073 1208 8

© The editors and contributors severally and the Publisher 2013

The authors' moral rights under the 1988 UK Copyright,
Designs and Patents Act are hereby expressly asserted.

All rights reserved. No part of this work may be copied, reproduced, stored,
sold, distributed, scanned, saved in any form of digital format or transmitted
in any form digitally, without the written permission of the Publisher.

BAR Publishing is the trading name of British Archaeological Reports (Oxford) Ltd.
British Archaeological Reports was first incorporated in 1974 to publish the BAR
Series, International and British. In 1992 Hadrian Books Ltd became part of the BAR
group. This volume was originally published by Archaeopress in conjunction with
British Archaeological Reports (Oxford) Ltd / Hadrian Books Ltd, the Series principal
publisher, in 2013. This present volume is published by BAR Publishing, 2016.

Printed in England

BAR titles are available from:

          BAR Publishing
          122 Banbury Rd, Oxford, OX2 7BP, UK
EMAIL   info@barpublishing.com
PHONE +44 (0)1865 310431
  FAX   +44 (0)1865 316916
          www.barpublishing.com

List of Contributors

SONIA FOCKE
Staatliches Museum Ägyptischer Kunst
Verwaltung
Arcisstraße 16
80333 München
DEUTSCHLAND

Email: sonia.focke@smaek.de

ALAN M. GREAVES
University of Liverpool
Archaeology, Classics and Egyptology
12-14 Abercromby Square
Liverpool, L69 7WZ
UNITED KINGDOM

Email: greaves@liv.ac.uk

BIRGITTA HOFFMAN
University of Liverpool
Archaeology, Classics and Egyptology
12-14 Abercromby Square
Liverpool, L69 7WZ
UNITED KINGDOM

Email: birgitta.hoffmann@liv.ac.uk

BARRY P. C. MOLLOY
University of Sheffield
Department of Archaeology
Northgate House
West Street
Sheffield, S1 4ET
UNITED KINGDOM

Email: barrymolloy@gmail.com

STEPHEN O'BRIEN
Department of History and Archaeology
University of Chester
Parkgate Road
Chester, CH1 4BJ
UNITED KINGDOM

Email: sobrien@chester.ac.uk

DAVIDE SALVO
State University of New York, Buffalo
Department of Classics
338 MFAC
Buffalo, NY 14261
UNITED STATES OF AMERICA

Email: davidesa@buffalo.edu

CAROLA VOGEL
Institut für Ägyptologie und Altorientalistik
Johannes Gutenberg-Universität
55099 Mainz
DEUTSCHLAND

Email: carolavogel@gmx.de

CONOR WHATELY
Department of Classics
University of Winnipeg
515 Portage Avenue
Winnipeg
Manitoba
R3B 2E9
CANADA

Email: c.whately@uwinnipeg.ca

Table of Contents

Introduction: Warfare and Society in the Ancient Eastern Mediterranean 1
Stephen O'Brien

'His Majesty Saw My Valour': Weapons as Rewards for Feats on the Battlefield.. 5
Sonia Focke

The *decimatio* in the Roman World.. 19
Davide Salvo

The Development of Warfare and Society in 'Mycenaean' Greece.. 25
Stephen O'Brien

Soldiers and Civilians – A New Look at Asymmetric Warfare in the Eastern Roman Empire in the 1st to 3rd century AD 43
Birgitta Hoffman

Militarization, or the Rise of a Distinct Military Culture? The East Roman Ruling Elite in the 6th century AD 49
Conor Whately

Malice in Wonderland: The Role of Warfare in 'Minoan' Society 59
Barry P. C. Molloy

Icon of Propaganda and Lethal Weapon: Further Remarks on the Late Bronze Age Sickle Sword................ 71
Carola Vogel

Post-Traumatic Stress Disorder (PTSD) in Ancient Greece: A Methodological Review 89
Alan M. Greaves

Introduction: Warfare and Society in the Ancient Eastern Mediterranean

Stephen O'Brien

Introduction

The study of violence and warfare is one of the oldest preoccupations of humankind, with early written sources often approaching the subject as myth or historical narrative. Attempts to comprehend human violence have been present in the humanities from the dawn of the discipline, and particularly in the work of the early modern philosophers Thomas Hobbes and Jean-Jacques Rousseau. Hobbes believed that humans in a 'state of nature' were invariably violent, with conflicting needs and desires resulting in the 'warre of every man against every man' (Hobbes 1651 [1991], 90). Rousseau (1754 [1984]), held that humans in a 'state of nature' would fight only for self-preservation, in the manner of any animal, and that it was the leaving of this state, with its attendant material inequalities, which ushered in other kinds of violence.

Debate as to the origins and causes of human violence has continued in various forms up to the present day, with sociobiologists and evolutionary psychologists often favouring explanations arising from the physical constitution of human beings (e.g. Wilson 1975; Pinker 2002), while those from the more humanistic social sciences tend to adopt positions in which human violence has largely cultural sources (e.g. Montagu 1976; Clastres 1994). While most contemporary researchers would accept that both genetic and social factors contribute to the genesis of violent acts, the precise weighting of these factors remains a matter of some dispute.

Several major historical developments of the past 25 years have served to both renew and inform academic interest in warfare and other forms of violence. The fall of the Soviet Union, for example, all but removed the threat of large-scale warfare from western society, yet contributed to a subsequent rise in ethnic conflict, exemplified by the Yugoslavian Wars of 1991-1999 and the Rwandan Genocide of 1994, which led to a new interest in questions about human violence (Vandkilde 2003, 136). Within archaeology and anthropology, these events are offered high-profile expression in Lawrence Keeley's *War Before Civilization: The Myth of the Peaceful Savage* (1996), a study which utilised ethnographic and archaeological data to contend that the human past has been unremittingly violent. This work has been heavily contested and its controversial nature has resulted in the generation of a substantial corpus of related literature. . Indeed, it is becoming increasingly difficult to keep track of studies of warfare, such is the pace at which they now appear, although significant examples include Jonathan Haas' *The Anthropology of War* (1990), Debra Martin and David Freyer's *Troubled Times: Violence and Warfare in the Past* (1997), John Carman and Anthony Harding's *Ancient Warfare: Archaeological Perspectives* (1999), Raymond Kelly's *Warless Societies and the Origin of War* (2000), Keith Otterbein's *How War Began* (2004), Mike Parker Pearson and Ian Thorpe's *Warfare, Violence and Slavery in Prehistory* (2005), Jean Guilaine and Jean Zammit's *The Origins of War: Violence in Prehistory* (2005), Elizabeth Arkush and Mark Allen's *The Archaeology of Warfare: Prehistories of Raiding and Conquest* (2006), and Ton Otto, Henrik Thrane and Helle Vandkilde's *Warfare and Society: Archaeological and Social Anthropological Perspectives* (2006). These volumes adopt a variety of theoretical perspectives on warfare, and provide a theoretical background against which we might examine the relationship between warfare and society.

Within the field of history, a gradual shift away from traditionalist battle- and campaign-centric 'drums and trumpets' military history has been apparent from the 1970s, largely replaced by the 'war and society' school with its greater emphasis on the cultural and historical context of warfare, and the technological emphasis of the 'military revolution' theory (see Brice and Roberts 2011; Wheeler 2011, for a fuller account of recent intellectual traditions in military history).

Both archaeologists and military historians have, therefore, endeavoured to contextualise warfare with reference to the broader themes of their respective disciplines, a necessary development given Parker Pearson's (2005, 21) recent assessment of the current state of the archaeology of warfare:

> Warfare has almost become an area of specialism in archaeology, formed out of the nexus of osteology, settlement studies and military history. This potential ghetto-ization is probably the very worst direction in which to go because the archaeology of violence is surely not to be relegated to a 'special interest' group in the same way as standing buildings or mummies. It has to be rescued from the wargamers and reinstated within the heart of a social archaeology. Warfare, violence and slavery are social institutions involving human agency within specific social contexts.

Given the often close relationship between history and archaeology apparent in the study of early Mediterranean societies, it seemed only correct that the research event which forms the basis of this volume serve as an interdisciplinary arena in which new perspectives from each might be discussed. The call for papers issued by the editors generated substantial interest, and particularly so from scholars at an early stage of their careers. The resulting conference encompassed a chronological span ranging from the Late Bronze Age to Late Antiquity, and a geographical area including south-eastern Europe, north Africa, and western Asia. This eclectic approach – positively received on the day – allowed researchers from diverse intellectual backgrounds and academic traditions to exchange perspectives in a manner which, we believe, can only ever be of benefit. . We hope that this volume captures that spirit of eclecticism, and allows these perspectives to be shared with an even wider and more diverse group of scholars.

About This Volume

Sonia Focke's *'His Majesty Saw My Valour': Weapons as Rewards for Feats on the Battlefield* analyses the possible practice of using weapons as reward for bravery during military campaigns in Old, Middle, and New Kingdom Egypt. Such a practice may explain the finds of weapons in private contexts which are inscribed with the names of both a ruler and a private individual. The significance of both the weapon types inscribed and the conventions of the inscriptions are explored, leading to analysis of the practices of weapon-ownership in Egypt during the periods in question, and considering whether weaponry was state property and distributed from arsenals, or the property of private individuals. The value of such practices of reward are made clear in the apparent equivalence of value between weapons and the 'gold of praise', the interment of weaponry with their recipients upon death, and the possibility that weapons retained as heirlooms conferred status between generations.

Davide Salvo's *The* decimatio *in the Roman World* analyses the Roman practice of decimation – the execution of randomly selected soldiers – as a form of military discipline. The paper adopts a diachronic approach, examining the practice as it developed between the Early Republican and Late Antique periods, and also examines its varying use in Europe and western Asia. Analysis suggests that the use of this method of punishment was not a constant of Roman military life, but saw periods of hiatus and revival. Its significance is made clear by examination of these historical patterns. While the intention of the practice no doubt be to ensure that soldiers fought bravely out of fear, and despite an evident comparison in the use of brutal public punishment to forestall civil rebellion, its hiatus after the Early Republican period instilled its subsequent reuse with a political dimension related to concepts of tradition and righteousness. The advent of the Christian era introduced a further social use to the practice, with martyrdom by decimation playing a role in early Christian thought.

Stephen O'Brien's *The Development of Warfare and Society in 'Mycenaean' Greece* seeks to reinterpret the Late Bronze Age of the Greek mainland by eschewing the social evolutionary concepts which have dominated the recent history of the discipline. Under this paradigm the Late Bronze Age societies of this region have been described as 'chiefdoms' which evolved into 'states' before collapsing to a 'simpler' form of society. The role of violence and warfare in this process was as a functional element allowing social and economic centralisation and hierarchy to be enacted. The paper therefore adopts a dual approach, presenting a view of Greek mainland societies which does not suggest progress on an evolutionary scale gauged by increases in hierarchy and centralisation, but rather one in which uneven and diffuse power changed historically. Similarly, warfare and violence are not treated simply as a functionalist tool, but rather as an element which could be deployed by any number of agents, in both ideological and material form, for an array of purposes.

Birgitta Hoffman's *Soldiers and Civilians – A New Look at Asymmetric Warfare in the Eastern Roman Empire in the 1st to 3rd century AD* moves away from a traditional focus on engagements between regular forces in the Roman Imperial period in order to examine the conflicts of Imperial Roman forces and irregular units. Examining a number of case studies provided by the Roman campaign in Numidia in AD 238, the paper examines the makeup of the *Legio III Augusta* – itself recruited in Numidia – to determine the extent to which identity would have been shared by the legionary forces and the irregular Numidian troops with whom they were fighting. Previous scholarly work has suggested that the massacres which took part during this campaign were the result of a social estrangement between the legionaries and the Numidian population, a consequence of the Roman army operating as a 'total institution' which replaced the previously existing identities and loyalties of its members. Hoffman's analysis, however, suggests that such social separations are not necessary for massacres to take place.

Conor Whately's *Militarization, or the Rise of a Distinct Military Culture? The East Roman Ruling Elite in the 6th century AD* uses a number of evidence types to challenge the theory that the aristocracy of the East Roman empire was unmilitarised until the mid-7th century AD, while the aristocracy of the post-Roman West was apparently militarised from the 3rd century AD onwards. The paper identifies militaristic ideologies in the textual and iconographic products of the East Roman empire, and notes that several neighbouring groups possessed elites commonly described as militarised. Following its exploration of the cultural aspects of militarisation in the East Roman empire, the paper then examines the social aspects of militarisation, by following the progression of individuals with military service into political or civilian state offices. A possible disjunction between the processes of cultural and social militarisation is identified, which opens up profitable new avenues of research.

Barry Molloy's *Malice in Wonderland: The Role of Warfare in 'Minoan' Society* tackles the subject of warfare in the Cretan Bronze Age, looking to dispel the false dichotomy between 'peaceful Minoans' and 'warlike Mycenaeans' which has characterised many previous treatments of the period. The paper assesses the full spectrum of available evidence for the Bronze Age practice of warfare on the island, demonstrating the numerous changes in both material culture and ideology which occurred over the two millennia in question. It is argued that when depositional differences are taken into account, both mainland and Cretan societies display a similar use of both martial values and technological systems. The frequent reliance in past accounts on invasive ethnic groups with different traditions of martial imagery and its display as an explanation of social change is thus rendered methodologically questionable. Warfare is therefore seen as an element which strongly structured the societies of the Aegean and played a significant role in the social dynamics of the period.

Carola Vogel's *Icon of Propaganda and Lethal Weapon: Further Remarks on the Late Bronze Age Sickle Sword* examines the presence of the *khepesh* or sickle sword in the Egyptian Late Bronze Age, taking advantage of the recent discovery of a number of new examples. . The paper first discusses the propagandistic use of the sickle sword in the art of the period, suggesting that the image of the weapon – often depicted in scenes of deities or ceremonial executions – was used to communicate ideas of both divine power and the physical prowess of the pharaoh. Following this, the paper examines the sickle sword as an item of material culture, utilising recent perspectives in the psychology of violence, and the technical properties of Egyptian weapons and armour, to show the increasing practical value of the weapon as the Late Bronze Age progressed. Both the social and practical aspects of the weapon are thus defined.

Alan Greaves' *Post-Traumatic Stress Disorder (PTSD) in Ancient Greece: A Methodological Review* critically assesses the recent body of research conducted on instances of Post-Traumatic Stress Disorder (PTSD) resulting from involvement in ancient warfare. The paper notes the difficulty in making individual diagnoses based only on ancient sources which took a different approach than our own to the writing of history, and further establishes that the shifting definition of PTSD in the modern era makes its identification in the past problematic. Indeed, the significant trans-cultural differences observable in the modern expression of psychological trauma suggests that current lists of symptoms may prove a poor guide for the identification of ancient examples. Ultimately, the paper does not suggest that scholars must take an agnostic view of the existence of PTSD in the ancient past, but rather that the literary portrayal of individuals changed by violence offers a window into the presence and cultural significance of trauma in a past society.

Acknowledgements

Our thanks are due to the Department of Archaeology, Classics and Egyptology at the University of Liverpool for the generous financial support which allowed the original colloquium to take place. All those who participated in the colloquium, whether as speakers, chairs, or audience members, played an enormous role in making the event a success. The authors of the individual papers have contributed their hard work and patience during the editing process, and it is their contributions that the volume rests on. Dr David Smith provided invaluable assistance with proof reading, while Dr Joseph Skinner and Tom McGrenery were able to suggest solutions to some of the thornier problems of formatting the manuscript for publication.

Bibliography

Arkush, E. N. and Allen, M. W. (eds.). 2006. *The Archaeology of Warfare: Prehistories of raiding and conquest*. Gainesville FL, University Press of Florida.

Brice, L. L. and Roberts, J. T. 2011. Introduction. In L. L. Brice and J. T. Roberts (eds.), *Recent Directions in the Military History of the Ancient World*. Publications of the Association of Ancient Historians 10, 1-10. Claremont CA, Regina Books.

Carman, J. and Harding, A. F. 1999. *Ancient Warfare: Archaeological perspectives*. Stroud, Sutton Publishing.

Clastres, P. 1994. *Archeology of Violence* (Translated by J. Herman). New York NY, Semiotext(e).

Guilaine, J. and Zammit, J. 2005. *The Origins of War: Violence in prehistory* (Translated by M. Hersey). Malden MA, Blackwell.

Haas, J. (ed.). 1990. *The Anthropology of War*. Cambridge, Cambridge University Press.

Hobbes, T. 1651 [1991]. *Leviathan*. Cambridge, Cambridge University Press.

Keeley, L. H. 1996. *War Before Civilization: The myth of the peaceful savage*. Oxford, Oxford University Press.

Kelly, R. C. 2000. *Warless Societies and the Origin of War*. Ann Arbor MI, University of Michigan Press.

Martin, D. L. and Frayer, D. W. (eds.). 1997. *Troubled Times: Violence and warfare in the past*. War and Society 3. New York NY, Routledge.

Montagu, A. 1976. *The Nature of Human Agression*. Oxford, Oxford University Press.

Otterbein, K. F. 2004. *How War Began*. Texas A&M University Anthropology Series 10. College Station TX, Texas A&M University Press.

Otto, T., Thrane, H. and Vandkilde, H. (eds.). 2006. *Warfare and Society: Archaeological and social anthropological perspectives*. Aarhus, Aarhus University Press.

Parker Pearson, M. 2005. Warfare, Violence and Slavery in Later Prehistory: An introduction. In M. Parker Pearson and I. J. N. Thorpe (eds.), *Warfare, Violence and Slavery in Prehistory*. British Archaeological Reports International Series 1374, 19-35. Oxford, BAR Publishing.

Parker Pearson, M. and Thorpe, I. J. N. (eds.). 2005. *Warfare, Violence and Slavery in Prehistory*. British Archaeological Reports International Series 1374. Oxford, BAR Publishing.

Pinker, S. 2002. *The Blank Slate: The modern denial of human nature*. London, Allen Lane.

Rousseau, J.-J. 1754 [1984]. *A Discourse on Inequality* (Translated by M. Cranston). London, Penguin.

Vandkilde, H. 2003. Commemorative Tales: Archaeological responses to modern myth, politics, and war. *World Archaeology* 35(1), 126-144.

Wheeler, E. L. 2011. Greece: Mad Hatters and March Hares. In L. L. Brice and J. T. Roberts (eds.), *Recent Directions in the Military History of the Ancient World* Publications of the Association of Ancient Historians 10, 53-104. Claremont CA, Regina Books.

Wilson, E. O. 1975. *Sociobiology: The new synthesis*. Cambridge MA, Harvard University Press.

'His Majesty Saw My Valour':
Weapons as Rewards for Feats on the Battlefield

Sonia Focke

Abstract

Autobiographical texts in Ancient Egyptian tombs often describe the rewarding of soldiers for feats of valour on the battlefield. They might receive the so-called 'Gold of Praise', or perhaps slaves captured in the battle; but also weapons, often described as wrought of precious metals. In this paper, some of the archaeological finds will be reviewed in the hopes of tracing these prizes and learning more about the history of some well-known objects, as well as addressing the issue of personal versus arsenal weapons.

Introduction

In the course of my doctoral research on Ancient Egyptian weapons, I have come across many pieces of weaponry inscribed with the name of a king. The impulse to attribute these weapons to the personal armament of the ruler in question is strong – yet further study jars with this impression. Leaving aside unprovenanced objects, some of these weapons were found in a clearly private context, and one even bears, in addition to the name of a ruler, the name of a private individual. Other objects seem too plain and of too poor workmanship to be dignified with the adjective 'royal'. To make the situation plain, we shall first give a brief overview of the material at hand (excluding inscribed axes from known foundation deposits).

Royal Weapons

In this category are weapons which, due to their archaeological context, materials and/or workmanship, can be considered as weapons actually borne by royalty.

Apart from the dagger of the Old Kingdom princess Ita, the first weapons that can be clearly attributed to the royal family from an archaeological context come from the caches of the 17th Dynasty queen Ahhotep and King Kamose. Kamose and Ahhotep each had bronze daggers, overlaid in gold and silver leaf, in opposite patterns: while Kamose's dagger has gold leaf on the pommel and blade, with silver on the handle, Ahhotep's has a gold handle with silver leaf on the pommel and blade. The fact that the blades themselves also have precious metals on them point to these being purely parade weapons, as opposed to weapons decorated with gold leaf in order to mark out the king on the battlefield (Petschel 2004, 131 Cat. No.131). A dagger of King Ahmose, found in his mother Ahhotep's tomb, has his name incised on the medial rib along with various desert scenes (Morenz 1999).

The tomb of Queen Ahhotep also contained several axes. One, with gold leaf and various inlays in decorative scenes and inscribed with the name of the king Ahmose, is obviously a royal weapon both by its materials and its inscription, as well as its archaeological context. The second, incised with the name of King Kamose, was unexceptional, and if not for the archaeological context, one would hesitate to term it a royal weapon (Kühnert-Eggebrecht 1969, 58 and our Figure 1).

Figure 1: Axe-head of Kamose (von Bissing 1900 Pl. III).

However, the tomb of Tutankhamun demonstrates that kings owned both simple and ornate weapons, and indeed that not all royal weapons bore the king's name (for example, some of Tutankhamun's self bows are quite simple and uninscribed) (McLeod 1982). In fact, that king had a marvellous array of different weapons: bows, both self and composite, two daggers (one of them with a blade of iron), two *khopesh*-swords, a suit of scale armour, four shields, a sling and various other pieces such as arrows and practice or duelling staves.

Weapons of Uncertain Archaeological Context

Unfortunately, most of the inscribed weapons were purchased on the art market or excavated at a time when provenance was a secondary consideration. A few examples are provided below to illustrate the wide variety of such blades.

One Type G axe in University College, London (Acc. No. 16324) bears the inscription 'Sesostris', the name of several well-known Middle Kingdom kings. However, the throne name is missing, and it seems likely that the blade dates to the Second Intermediate Period (which also brought forth kings of that name), as this type is known from the Pan Graves of the Second Intermediate Period, as well as from the 18th Dynasty (Kühnert-Eggebrecht 1969, 29 ff.). It is equally possible that his axe had an amuletic function, possibly akin to the openwork axes discussed later on (Kühnert-Eggebrecht 1969, 58).

Another axe, in the British Museum (BM 67505), bears the name of a king called Djedu-ankh-Re, possibly of the 13th Dynasty, while another mentions the Second Intermediate Period king Nebmaatre (BM 63224).

A dagger with the cartouche of Apophis Aaqenenre ('the Good God Aaqenenre') was purchased by H.A. Corble in Luxor (Figure 2). The royal name was incised onto the blade rather than inlaid (Dawson 1925, 216).

Figure 2: Dagger with the cartouche of Apophis Aaqenenre (Dawson 1925, Pl. XXV).

In Luxor, Golenisheff bought a lance head, now in the Pushkin Museum of Fine Arts in Moscow (Acc. No. I.1.a.1762) inscribed with the names of Ahmosis and the following inscription:

ntr nfr (Nb-pḥtj-Rʿ) sꜣ Rʿ (Jʿḥ-ms) dj ʿnḫ jnt n.f m nḫtwt.f m ḥwt-wʿrt ḥsjt

'The Good God Nebpehtire, Son of Re Ahmosis, may he be given life. Bringing back by him from his victories in Avaris the miserable.'

The inscription makes it clear that this piece was booty, reworked to bear the name of the king who acquired it (Hodjache and Berlev 1977).

An axe from the British Museum (BM 36770) and, due to its size, probably not from a foundation deposit, is incised with the throne name of Tuthmosis III.

The Egyptian Museum in Florence owns a dagger (Acc. Nr. 7677) bought in Luxor by Schiaparelli (Figure 3), and thus of uncertain provenance. It has an unusual pommel made up of two hawk heads; the grip, of plain wood, is incised with the throne name of Tuthmosis IV. The workmanship is good but not remarkable.

Figure 3: Dagger Florence 7677 with cartouche of Tuthmosis IV (reproduced courtesy of the Museo Egizie, Firenze).

Bearing the name of his predecessor Sethi I is an inlay for the handle of a *khopesh*-sword from the Brooklyn museum (Acc. Nr. 49.167), of niello with copper and gold inlays (Figure 4). The precious metals and technical excellence may point to it being the weapon of a king (Müller 1987, 153).

Furthermore, a *khopesh*-sword bearing the names of Ramses II survives (Louvre E 25689) (Figure 5) – however, the inscription seems rather carelessly incised, which may point to a hasty execution before this simple weapon was donated to its new owner (see below) (Müller 1987, 152).

One bronze long-sword in the Egyptian Museum in Berlin (Acc. Nr. 20305) bears both the birth and the throne name of Sethi II in gold inlay. Other than this inlay, nothing points to its being a royal weapon;

Figure 4: *Khopesh* handle inlay with the names of Sethi I (Müller 1987, Pl. XVII).

Figure 5: Handle of Louvre E 25689 with the name of Ramses II (copyright Musée du Louvre/Maurice et Pierre Chuzeville).

however, this type of bronze sword, in use in Central Europe and some parts of the Mediterranean, is known from only four examples in Egypt, so it seems likely that it was a royal weapon, possibly a diplomatic gift (Burchardt 1912; Petschel 2004, 136).

An axe from the Third Intermediate Period and now in the British Museum (BM 66211) is inscribed with the name of a king named Usermaatre-Setepenre. However, the inscription is engraved, a technique which is unlikely to have existed in the Third Intermediate Period, so it seems likely that, while the axe is genuine, the inscription may not be (Davies 1987, 47).

Weapons Bearing Royal Names from a Private Context

Not all weapons bearing royal names should be attributed to a royal context, however. A dagger bearing the name of the king Se-wadj-en-Re (*sw3d-n-r‛*) of the 14th Dynasty was found by Petrie in a private tomb context in Diospolis Parva (Figure 6); it is a very fine example of the typical Middle Kingdom / early Second Intermediate Period type, with the rivets of the handle inlays adorned with rosettes (Kühnert Eggebrecht 1969, 58 and Petrie 1901, 52 and Pl. XXXII).

The inlay for a dagger handle, bearing both names of Tuthmosis I, comes from the tomb of the early 18th Dynasty soldier Ahmose-Pen-hat, southwest of Deir-el Bahari (unfinished mortuary temple of Seankh-ka-re Mentuhotep), and is now in the Metropolitan Museum of Art in New York (Hayes 1959, 76-77 and fig.40) (Figure 7).

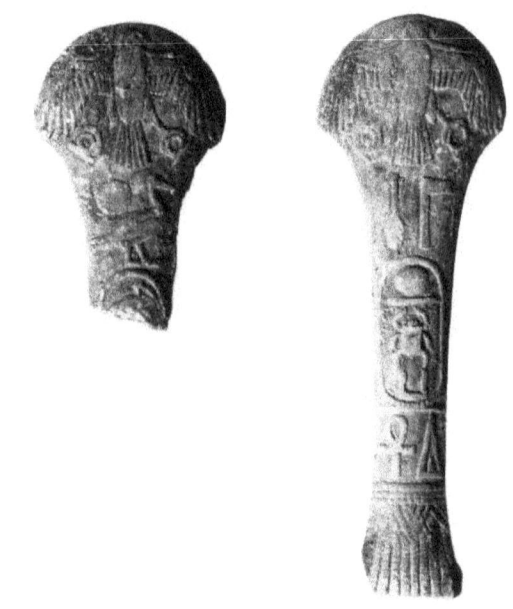

Figure 7: Dagger handle inlay of Ahmose Pen-hat (Hayes 1959, fig. 40).

These prove that a royal name need not necessarily grace only royal weapons. What these side arms were doing in private graves, however, only becomes clear when we observe a fascinating group of weapons inscribed not only with royal but also with private names.

One interesting dagger from the Second Intermediate Period was found in a private tomb dug within the funerary enclosure of the Old Kingdom queen Apuit in Saqqara. It is made of bronze, with a wood inlay handle overlaid with electrum and bearing a hunting scene, the name of the 'Good God, Lord of the Two Lands Neb-nem-Re Apep, may he live' and that of a 'Follower of his Lord, Nehemen' (Daressy 1906) (Figure 8). This is of interest, because the owner of the tomb appears to be an individual named 'Abd...' – unless 'abd' should be considered a designation of lineage. This makes it abundantly clear that, while a very precious piece and bearing a royal name, it was the property of a private person.

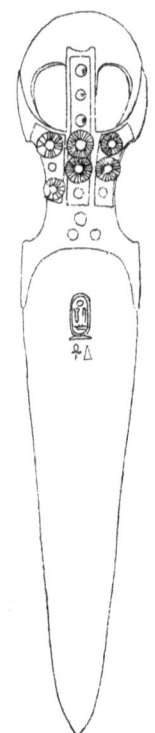

Figure 6: Dagger from Diospolis Parva with name of Se-wadj-en-Re (Petrie 1901, Pl. XXXII).

What is more, there exist three other axes bearing the name of Kamose and which, according to Kühnert-Eggebrecht (1969, 58), may have come not from his tomb, but from those of private individuals. They are now in Oxford (Ashmolean 1927/4623), London (BM 36772), and Cairo (CG 52648).

An axe inscribed with the name of the Middle Egyptian Hyksos vassal Neb-maat-Re was found in a Pan-grave burial in Mostaggeda. Kühnert-Eggebrecht (1969, 58) believes this piece was probably awarded from the booty taken from the enemy. Unfortunately, however, the plates she mentions show only uninscribed daggers of the Middle Kingdom type.

Figure 8: Dagger of Nehmen, now in the Luxor Museum (Daressy 1906, Fig. 1 and 2).

Schmitz (1977, 215-216) mentions a throwstick, also from the Second Intermediate Period, found by Mariette in the tomb of Ak-Hor in Thebes (Figure 9), bearing both the royal name 'Taa' (either Senakhtenre or Sekenenre) and that of the *s3-nswt* Tjuiu, who is said to 'accompany (*šms*) his lord on his paths.' The King's Son Tjuiu need not be an actual prince, as this honorary title was often bestowed on fortress commanders (*ṯsw*) in the Second Intermediate Period. However, Schmitz's identification of the object as a throwstick is somewhat puzzling as, according to Mariette's plates, the object measures some 1.25m in length. Its width seems to preclude it being a bow, so that it seems most likely that it was a stave of some sort – staves and walking-sticks also being attested as gifts from the king both in textual and archaeological evidence (Schmitz 1977, 213-214).

A rather extraordinary bow is now in the Pushkin Museum of Fine Arts in Moscow (Acc. Nr. I.1.a 1804) (Figure 10). It bears the name of the Second Intermediate period Theban king Rahotep, with the following inscription:

Nswt (Rˁ-ḥtp) wḏ n s3-nswt Jmnjj jm rdj.tw t3 pḏt ḥr šms Mjn m ḥb.f nb s3 n s3 n jwn n jwn

Berlev believes it was given to the Prince Ameni to be used by him and his descendants for the feasts of Min – so it appears to be not a military weapon, but a ritual one; unless it was meant to accompany the god's procession in the hands of guards or soldiers (Berlev 1975/76). Schmitz (1977, 216-218), though, believes that not one, but two private names appear on the bow: the prince Ameni and the actual recipient of the bow, a certain Minemhebef-neb.

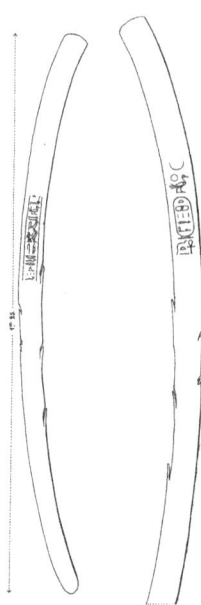

Figure 9: The 'throwstick' of Ak-Hor (Mariette 1889, Pl. 51).

Rather than translating the text as: 'the king Rahotep ordered that be given to the Prince Ameni this bow to

Figure 10: Bow in the Pushkin Museum Acc. Nr. I.1.a 1804 (Berlev 1975/76, Pl. II).

accompany Min in all his festivals' Schmitz reads it:

'Order of the King Rahotep to the Prince Ameni: let this bow be given to the officer (šms) Minemhebef-neb, from son to son, heir to heir.' If this is so, we have a clear case of a weapon given to a soldier from the king – this time through the offices of a prince or fortress commander.

The honour given is hereditary, and the bow is meant to be passed on to the man's heirs, whether that man be Ameni or Minemhebef-neb.

This is not the only weapon bearing mention of its nature as a gift. Another such, an axe-head now in the British Museum (BM 37447) (Figure 11), bears not only the name of the king, but an inscription detailing that it was a reward (Figure 12).

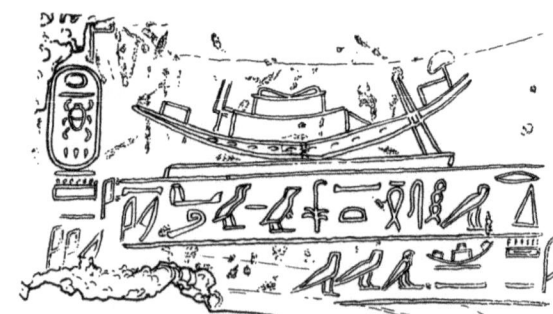

Figure 12: Inscription on the axe BM 37447 (Davies 1987, Pl. 32,126. Reproduced courtesy of the British Museum).

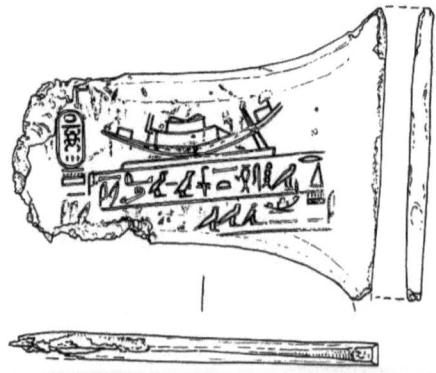

Figure 11: BM 37447 (Davies, Pl. 7, 156. Reproduced courtesy of the British Museum).

rdj m ḥsj nswt n wʿw n Mrj-Jmn n Nḥ-mm

'Reward from the king given to the sailor of the ship *Beloved-of-Amun*, Nehmem', together with the cartouche of the king Amenhotep II (Kühnert-Eggebrecht 1969, 59-60 and Davies 1987, 156), whereby the reading of the *w'w*'s name remains problematical. Following Edwards, the bird appears, to my eyes, nearer to the *Ḥr*-Falcon than anything else, but has been read variously as *nḥ* (Davies) or *3* (Kühnert-Eggebrecht). The reading *3mm* suggests the possibility of a lexicographical error for the name *'mm* of Ranke's list.

However his name is to be read, it is clear from the inscription that the soldier from the *Meri-Amun* was awarded this axe by the king, and probably had the inscription added himself, where the careful placement of the king's name suggests that the name of the ship was not merely *Beloved of Amun*, but should be read *Amenhotep Beloved of Amun* instead (Kühnert-Eggebrecht 1969, 60).

Another, earlier piece also hints at this practice. BM 20923, the butt end of an axe shaft, is inscribed with the name of 'the Good God, Sekhemreswadjtawy', the 13th Dynasty king Sobekhotep III (Figure 13). The fragment also retains the first part of an inscription going the length of the haft and starting with the word *rdj*, 'to give' or 'given'.

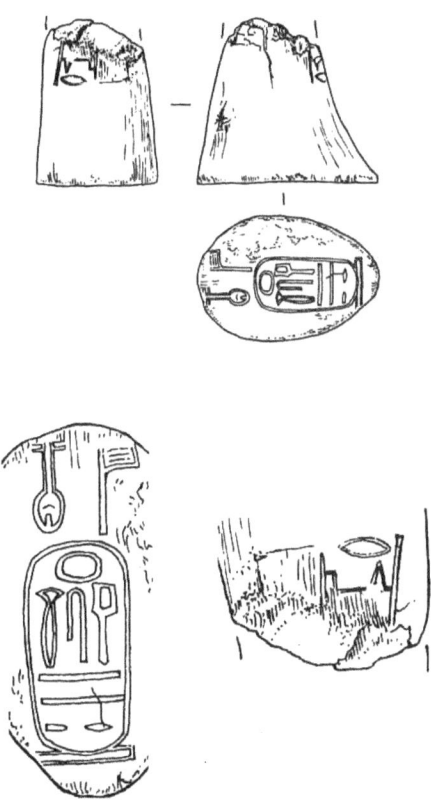

Figure 13: Axe shaft fragment BM 20923 with detail of inscription (Davies 1987, Pl. 30 and 31. Reproduced courtesy of the British Museum).

This practice appears to have continued into the Late Period, with an axe naming both 'the Great Chief of the Libyans, Uasarharta' and 'Horsiese, son of Hor-shenit' (Kühnert-Eggebrecht 1969, 60):

These objects prompted a more acute look into the autobiographical inscriptions in Ancient Egyptian private tombs. The granting of rewards from the king to a private individual in terms of gold and other precious goods is a well-known occurrence; it is also known that this happened in a military context as rewards for special feats on the battlefield, such as the capture of prisoners or killings ('taking of hands' – downed enemies had their hands cut off for tallying purposes): 'Meanwhile I was at the head of our troops, and His Majesty saw my valour. I brought off a chariot, its horses, and him who was upon it as a living prisoner, and took them to His Majesty. One presented me with gold in double measure' (after Breasted 1988, 34-35).

A careful review of the records shows that, apart from gold in the shape of jewellery such as the necklaces and armlets/bracelets that usually make up the 'Gold of Praise' (*nwb n ḥswt*; attested since the 6th Dynasty) or the little flies and lions that were purely military honours, weapons, too, were given away as rewards (Butterweck-Abdel Rahim 1999, 234).

This practice is not documented for the Old Kingdom, where rewards were mainly of a funerary nature (Butterweck-Abdel Rahim 1999, 32), but appears for the first time in written records of the Middle Kingdom, about when the designation 'Gold of Valour' (*nwb n ḳn.t*) comes into use, though it does not seem to strictly designate gold flies or lions, but simply Gold of Praise given for military valour as opposed to political favour. In fact, von Deines (1954, 86) argues that in this case the *n* does not designate an indirect genitive, but rather the preposition 'because, due to,' so that *nwb n ḳn.t* should be translated 'gold because of valour'.

Khu-u-Sobek, whose stela from Abydos is now in the Manchester Museum (acc. nr. 3306), receives for his bravery in the Syrian campaign of Sesostris III a *sṯs*-staff, a bow (*jwn.t*)' and 'a dagger (*mḳsw*) worked in electrum' (Butterweck-Abdel Rahim 1999, 54 and Goedicke 1998, 36):

ḥn' n rdj.n.f n.j sṯs m w3s m nwb rdj.tw.j jwnt ḥn' mgsw b3k m w3s m nwb ḥn' ḥ'wwf

'And he gave to me a staff of electrum; I was (also) given a bow, together with a dagger worked in electrum together with his (other?) weapons.'

Goedicke (1998, 36) reads the last term as *ḥnꜥ ḫfꜥw*, which Goedicke believes either denotes a technical feature of the weapon, or else is derived from **ḥnꜥy ḫftjw*, 'belonging to the enemy', suggesting that the weapon was booty. It could also derive from *ḥf*, 'plunder', or *ḥfꜥ*, 'grasp, take' or 'to make booty' (Hannig 1995, 597). However, the use of *ḥnꜥ*, 'together with' suggests that the word should, in fact, be read exactly as it is written, *ḫꜥww.f*, meaning 'his weapons'. Either way, a redistribution of the weapons taken from the defeated enemy (assuming that is who is meant by 'his') is suggested. This was one of the oldest practices for the reward of soldiers; the division of booty is common in many cultures and occasionally even set down in law (as in the Laws of Hywel Dda in medieval Wales).

As evidenced from his Edfu stela, the *s3-nswt m ḥḳ3 n nḫt* Hor-sekher received a golden axe (*ꜥkh*) from the hands of a Second Intermediate Period king called 'Dedu-mesu' (Schmitz 1977, 214).

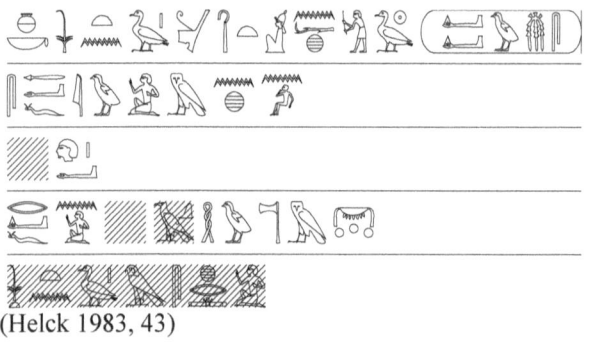

(Helck 1983, 43)

Jnk s3-nswt n ḥḳ3t nḫt s3-Rꜥ (*djdjw-ms*)
sꜥ3.f jw m nḫn
...*tp ꜥ*
rdj.f n.j ... 3ḫḥw m nwb
s3-nswt ḥr-sḫr

'I am a King's Son of the victorious ruler, the Son of Re (Dedu-Mesu)
He made me great as a child
….. at his hand.
He gave to me an axe in gold,
(to me,) the King's Son Hor-sekher.'

This practice survived into the New Kingdom; Ahmose-pen-Nekhbet reports in his tomb in El Kab (after Sethe 1963, IV 38-39):

'King Djeserkare (Amenhotep I), triumphant, gave to me, of gold: two bracelets, two necklaces, an armlet, a dagger, a headdress, a fan, and an arm ring.

King Aakheperkare (Tuthmosis I), triumphant, gave to me, of gold: two bracelets, four necklaces, one armlet, six flies, three lions, two golden axes.

King Aakheperenre (Tuthmosis II), triumphant, gave to me, of gold: three bracelets, six necklaces, three armlets, an arm ring; a silver axe.' (after Breasted 1988, 11)

According to the depictions in his tomb, the Steward and Fanbearer Tjenuna, who served under Tuthmosis IV and Amenophis III (TT 76), may have received not only statues, pectorals, a mirror, *vbjw*-chains, a censer-arm, a *msktw*-armlet, a *mnfrt* made from beads and two *ꜥwꜥw*-bracelets, but also a mace and a *khopesh*-sword (Butterweck-Abdel Rahim 1999, 127-128). However, though Butterweck-Abdel Rahim chooses to interpret this scene thus, it is equally possible that the gifts were going the other way, and were meant, instead, for the king, either as New Year's gift or presentation of booty (Säve-Söderbergh 1957, Pl. LXXII). Unfortunately, the tomb decoration is very damaged, and the surviving text is very uninformative in that regard.

After Tjenuna, no more mention is made of weapons as rewards (Butterweck-Abdel Rahim 1999, 217), although, as we have seen above, the practice appears to have survived into the Libyan Period.

Other types of weapon that may have been received as rewards are those with purely ceremonial function, not meant for battle at all yet still implying it by their form.

Weapons Bearing No Inscription But Not Meant for Battle

From the Middle and New Kingdoms, axes have been found bearing openwork decoration. As this would greatly weaken the axehead, it seems likely that they had a purely ceremonial function – perhaps as parade weapons given to particularly deserving soldiers. They mostly bear scenes of apotropaeic function: figures of gods, Djed-pillars, but also hunting scenes, or a bullfight (Kühnert-Eggebrecht 1969, 78), signifying victory over the forces of chaos and recalling the hunting scene on the dagger of Nehemen. Such hunting scenes may have had an amuletic quality more specifically geared towards the soldier. Unfortunately, neither texts (the axeheads themselves are never inscribed) nor the archaeological context, usually unknown, of these openwork axeheads can confirm this.

These axeheads are all of Kühnert-Eggebrecht's (1969, 62-91) Type D or Type G IX. G IX dates to the New Kingdom; D is somewhat ambiguous, but seems to span the early Middle Kingdom into the Second Intermediate Period. Though Type G IX is a war axe, Type D is known mostly as a tool. However, the latter was possibly chosen for displaying openwork decoration because the war axes of the time, with their long slim blades (Type E and its variants), were not adapted for it.

It is probably for the same amuletic purpose that one full blade was inscribed with the name of an unidentified Sesostris, and two others were adorned with scratched-in depictions of Thoeris – a common motif of openwork axes – and a falcon-headed god (possibly Monthu, or Re) respectively (Kühnert-Eggebrecht 1969, 58-59).

Preliminary Conclusions

All in all, we have eight, possibly nine examples in which the gift of a weapon by the king to a private person is likely (Table 1).

Despite the fact that ten examples do not a statistically viable basis make, it is interesting to see that the bow only appears in the Middle Kingdom and the *khopesh*-sword not before the late 18th Dynasty – curiously enough, together with the mace, which was, as far as can be ascertained, no longer used in actual combat at that time.

By far the most common weapon accorded is the axe, followed by the dagger and the bow. Throwstick, *khopesh*-sword and mace each appear only once and in uncertain context.

Also of interest is the designation 'Follower of his Lord' and its variations, which seem to be linked with the giving of weapons:

- Nehemen is a 'Follower of his Lord'.
- Tjuiu 'follows his lord on his paths'.
- Minemhebef-neb, if he exists, is a 'Follower'.
- Djehuti is also termed 'Follower of his Lord'.

The question arises, of course, whether these men were 'Followers of their Lord', i.e. belonged in the King's train (*vmsw*), *before* they received the weapon, or if the weapon was, among other things, a sign of the granting of the honorary title *vmsw*.

The granting of (honorary) titles or epithets for military prowess is not unknown. Among such titles are: *kf‘w*, 'Raider, Booty-maker' (Hannig 1995, 882); *knj n nswt*, 'Brave of the King' (Hannig 1995, 858), borne by Ahmose-Pen-Nekhbet; and *‘h3wtj n hk3*, 'Fighter/Warrior of the Ruler'(Hannig 1995, 153), a title borne by General Djehuti; perhaps 'Follower' can be added to them?

Most of the recipients of weapons from the king have military titles, though Nehemen and Minemhebef-neb are both simply 'Followers', a title/epithet which may or may not have a military connotation, and Tjenuna is Steward and Fanbearer – in which capacity he may have followed the king into battle, however.

In some cases, the king's name appears to have been added to these weapons as a reminder of the honour received – either by the king himself (or rather his artisans), as is probably the case for the gilded hilt of Nehemen's dagger, or possibly incised by the recipient (or in his name by an artisan or scribe) as a way of showing off. It is also probable that weapons such as the *khopesh* with the name of Ramses II, where the inscription is hastily scrawled, belonged not to the king himself, but to a brave soldier who received it as a reward.

Privately Owned Weapons?

The distribution of these weapons to deserving soldiers and their subsequent appearance in tombs raises the question of the private ownership of weapons, and with it that of arsenals. It is still unclear which weapons were arsenal-stored and distributed to the soldiers by the state, and which they were expected to provide themselves. It is not unknown for soldiers, especially officers, to have to provide some or all of their equipment. Medieval knights had to provide their own equipment (Keen 2002, 45); in Southern cavalry units of the American Civil War, horses were private property and had to be furnished and replaced by the cavalry soldiers themselves (Boger 1995, 96-97). Were the weapons gifted to soldiers as a reward, then, arms that were otherwise also privately owned, or were they standard-issue weapons that passed into private ownership?

Compared with axes and especially, spears, daggers seem to prevail in the archaeological evidence as tomb offerings (Vila 1970, 191). Model spears and shields were occasionally present in tombs, but functional examples appear to be comparatively rare. Axes seem to appear more seldom than daggers, though this would have to be confirmed statistically. Bows, though common, are a difficult case, as there is no way to distinguish hunting weapons from military ones. One is therefore inclined to see daggers as personal weapons, which could be taken to the grave, as opposed to arsenal-issued spears, shields and axes which would have to be returned after battle or upon leaving the army. Indeed, daggers are rarely depicted in battle scenes, making it likely that they were not mass-issued. The same is true of *khopesh*-swords, though by the Ramesside Period whole platoons are armed with them, rather than being scattered amongst individual soldiers, so it seems probable that, by this time at least, they had become standard issue.

However, other explanations exist for the predominance of daggers in the archaeological evidence: they may have been status symbols, and as such acquired by persons with no military connections, or they may also have been officially issued upon attaining a certain rank, and then regarded as personal property.

We know, for example, that a private individual could acquire a chariot for the training of war horses (Hofmann 1989, 68). What is not clear is whether the chariot was considered his personal property or that of the Crown. Furthermore, the question arises as to whether, in time of war, one had to supply one's own chariot or could expect one to be issued. That chariots were repaired on site is known from the Kadesh camps depictions, but whose responsibility it was to ensure their upkeep is not clear.

A simple answer to this question could be ascertained if more was known of Ancient Egyptian arsenals. However, the evidence is slight and scattered throughout Egyptian history.

For the Old Kingdom, the *pr-ꜥḥꜣ*, or 'Weapons-House', probably denotes the central arsenal. Helck mentions that the arsenal was probably part of the Treasury administration, as 'Overseers of the Weapons-House' were often also 'Overseers of the Treasury' (Helck 1954, 65). In the tomb of Senebi in Elephantine, the *pr-ꜥḥꜣ* is also mentioned next to the *pr-hd* (Sethe, 1903, 138,6) as a source of objects for burial.

In the New Kingdom, the word *ḫpv* is considered to be the word for arsenal. It is used in the pAnastasi I 26, 3-5 to designate the weapons smithy where a charioteer returning from Palestine goes to have his chariot repaired (Herold 2003, 197). In the New Kingdom, *ḫpv* appears twelve times in the context of weapons manufacture, including its mention in the reliefs of Ki-iri. It seems to encompass both the weapons-smithies and the storage facilities for the finished product. Extant descriptions and depictions of armouries suggest that workshops and magazines were grouped around a central courtyard with a single entrance. The *ḫpv* appears also to have been under the aegis of the Treasury (Herold 2003, 199).

Fortunately, we do have an archaeological record of the arsenal of the fortress of Mirgissa, an Egyptian outpost of the 2nd millennium BC at the second cataract in Nubia – though it is unknown if the entire armoury was preserved or only one section. Remains of spears, lances, bows, arrows, several axes and the facilities for making and repairing shields were uncovered (Vila 1970). The absence of daggers could point to their not being arsenal weapons – or else they were kept in another, less well preserved part of the fortress.

Finding depictions of actual arsenals is difficult. Many reliefs depicting weapons show only the workshops, or New Year's gifts for the king, or else reflect the tomb owner's wish to face the afterlife well-armed.

The best-known of these representations is the so-called 'arsenal' of Ramses IV. In one of the side chambers in his tomb, the walls were decorated with representations of weapons and various divine standards (Rosselini 1834, PL. CXXXI). The decoration of other side chambers from the same tomb, showing pieces of furniture and statues of the king, tells us that, as in some temple magazines, they show the content of that chamber. Therefore, the depictions of bows, arrows, quivers, shields, *khopesh*-swords, daggers, scale armour, whips, spears, helmets, axes and some form of single-edged blade depict, not necessarily the true content of an arsenal – though it resembles the content of the 'arsenal' of Ki-iri (see below) – but rather the dead king's armament for the afterlife.

Of particular interest is the 'arsenal' of the tomb of Ki-iri from Saqqara. Unfortunately, not all the blocks, now in the Cairo Museum (JE 43275 and further blocks in the temporary register), survive; however, there are enough to get a general overview of the scene. Ki-iri, 'Head of the Chariot Artisans' (*hrj hmww wrrjt*), 'Head of the Chariot Artisans of the Armoury' (*hrj hmww wrrjt n pꜣ-ḫpv*) and 'Overseer of the Artisans of the Lord of the Two Lands' (*jmj-r hmwt n nb-tꜣwj*) (Herold 2003, 196), sits on a chair while overseeing the various weapons-producing workshops. Several workers bring the finished products to him and lay them at his feet. Behind him, storerooms are filled with weapons. In my original paper I maintained that no daggers are present in the series of rooms depicted behind Ki-iri; however, since then I have come across a larger depiction of the scene (Herold 2006, 72-73) and realised that what I had interpreted as spears were actually coats of scale armour with, before them, tables laden with daggers. The storage rooms seem to contain all the types of weapons available to the Egyptian military and produced by the armoury workshops.

Unfortunately, this still fails to answer the question of personal weapons. The storerooms could comprise the military arsenal, or simply represent the storerooms for the workshop, their contents then being either sold separately or distributed to the soldiery. For example, chariots are depicted, and, as mentioned above, we know that it was possible for a private person to purchase a chariot and chariot parts in order to train a team of horses (Hofmann 1989, 68).

The most interesting scene for the question of military-issue weaponry is a scene from Ramses III's temple at Medinet Habu (Breasted 1930, Pl.29) (Figure 14).

Before setting out against the Sea People, the king issues equipment to his troops, while the scribes see to the distribution (Kitchen 1983, 28):

rdjt ḫꜥw n mšꜥ n nt- ḥtrj n pḏwt šꜣrdnw nḥsjw

'Issuing weapons to the infantry and the chariotry, and the troops of Sherden and Nubians.' (Kitchen 2008, 25)

It is clear from this text that all areas of the military are being supplied: infantry, chariotry and foreign mercenaries. The equipment depicted being handed out encompasses helmets, spears, composite bows, quivers (presumably filled with arrows), *khopesh*-swords and oddly-shaped implements that may represent sheaths for the swords (see Förstner-Müller 2007-2008 for the existence of sheaths for *ḫpv*-swords). This is the only document which clearly shows what equipment was standard troop issue, at least during the reign of Ramses III. Axes (assuming we are correct in believing them to be standard issue in earlier times) are conspicuous by their absence, apparently replaced by *khopesh*-swords. Present in the scene from the tomb of Ki-iri, but not here, are daggers.

Figure 14: Ramses III distributing weapons to his soldiers.

Conclusions

Though there is still much to be done in the study of arsenals and the personal ownership of weapons, a few preliminary conclusions in regard to private weapons inscribed with the name of a king may be allowed.

A great many different types of weapon were awarded for feats of valour: axes, daggers, bows, *khopesh*-swords and even a mace. The gifting of arms seems to have been on a par with the awarding of the 'gold of valour', which could also take the form of jewellery. It also appears to be connected with the title of 'Follower (of his Lord)'. The weapons, once awarded, were of course considered the soldier's personal property, but need not be of the type usually reserved for personal ownership.

Daggers, for example, seem to have been manufactured in the royal armouries, but appear not to have been distributed to soldiers en masse, as they are apparently absent from the arsenal at Mirgissa and do not appear in the distribution scene of Medinet Habu. However, as they do occasionally appear in the hands of individual soldiers in battle-scenes, they must have been acquired by these men in some fashion. The author favours an interpretation of daggers as personal weapons a well-to-do soldier or officer could purchase from the armoury, though the thesis that they were weapons only handed out once a soldier had achieved a certain rank also has merit. Be that as it may, it seems that, since daggers were taken to the grave, they were considered personal property. They were also common enough gifts from the king to deserving military underlings.

Other weapons given as gifts, though, seem to have been of the general-issue type. Bows are attested both at Mirgissa and in the distribution scenes of Medinet Habu, and axes at least seem to be comparatively rarer than daggers in tombs (though this needs to be statistically verified and may vary from period to period) and are depicted in the hands of whole divisions of infantry soldiers, which would argue for their being mass-issue weapons returned after use. *Khopesh*-swords seem to have started as personal weapons, only to become generalised in the Ramesside Period. The only anomaly is the mace given to Tjenuna, which is not attested as a weapon of war after the Predynastic Period.

It seems that the gifting of an otherwise general-issue weapon to a soldier as a reward for bravery on the battlefield earned him the right to consider it his own private weapon, which he could then take with him into the Afterlife. Indeed, the receipt of a handsome new weapon (or three) from the 'Living God' himself must surely have spurred a valorous soldier on to further acts of bravery on the king's behalf, as well as elevating his status – both in this life and the next.

Bibliography

Berlev, O. 1975/1976. Un don du roi Rahotep. *Orientalia Lovaniensia Periodica* 6/7, 31-41.

Boger, J. 1995. *Der US-Bürgerkrieg 1861-1865. Soldaten – Waffen – Ausrüstung*. Stuttgart, Motorbuch Verlag.

Breasted, J. H. (ed.) 1930. *Medinet Habu Volume I Earlier Historical Records of Ramses III*. University of Chicago Oriental Institute Publications Vol. VIII, Chicago, University of Chicago Press.

Breasted, J. H. 1988. *Ancient Records of Egypt, Vol. II: The Eighteenth Dynasty*. London, Mysteries of Man Ltd.

Burchardt, M. 1912. Zwei Bronzeschwerter aus Ägypten. *Zeitschrift für Ägyptische Sprache und Altertumskunde* 50, 61-63.

Butterweck-Abdel Rahim, K. 1999. *Untersuchungen zur Ehrung verdienter Beamter*. Aegyptiaca Monasteriensia 3. Münster, Shaker Verlag.

Daressy, G. 1906. Un poignard du temps des rois pasteurs. A*nnales du Service des Antiquités Égyptiennes* 7, 115-120.

Davies, W. V. 1987. *Catalogue of the Egyptian Antiquities in the British Museum VII, Tools and Weapons I, Axes*. London, British Museum Publications.

Dawson, W. R. 1925. A Bronze Dagger of the Hyksos Period. *Journal of Egyptian Archaeology* 11, 216-217.

Forstner-Müller, I. 2007-2008. A new scimitar from Tell el-Dab'a. *Archaeology and History in Lebanon* 26-27, 207-211.

Goedicke, H. 1998. Khu-u-Sobek's Fight in 'Asia'. *Ägypten & Levante* VII, 34-37.

Hannig, R. 1995. *Großes Handwörterbuch Ägyptisch-Deutsch*. Mainz, Philipp von Zabern.

Hayes, W. C. 1959. *Scepter of Egypt Vol. II the Hyksos period and the New Kingdom 1675-1080 B.C.* New York City, Metropolitan Museum of Art.

Helck, W. 1954. *Untersuchungen zu den Beamtentiteln des ägyptischen Alten Reiches*. Ägyptologische Forschungen 18. Glückstadt, J. J. Augustin.

Helck, W. 1983. *Historisch-biographische Texte der 2. Zwischenzeit*. Wiesbaden, Otto Harrassowitz.

Herold, A. 2003. Ein Puzzle mit zehn Teilen – Waffenkammer und Werkstatt aus dem Grab des Ky-jrj in Saqqara. In N. Kloth, K. Martin, and E. Pardey (eds.), *Es werde niedergelegt als Schriftstück. Festschrift für Hartwig Altenmüller zum 65*. Beiträge zu den Studien zur Altägyptischen Kultur 9, 193-202. Hamburg, Buske.

Herold, A. 2006. *Streitwagentechnologie in der Ramses-Stadt. Knäufe, Knöpfe und Scheiben aus Stein*. Die Grabungen des Pelizaeus-Museum Hildesheim in Qantir-Piramesse, Band 3. Mainz, Philipp von Zabern.

Hodajache, S. and Berlev, O. 1977. Objets royaux du Musée des Beaux-arts Pouchkine à Moscou. *Chronique d'Égypte* 52, 22-37.

Hofmann, U. 1989. *Fuhrwesen und Pferdehaltung im alten Ägypten*. Inaugural-Dissertation zur Erlangung der Doktorwürde der Philosophischen Fakultät der Rheinischen Friedrich-Wilhems-Universität zu Bonn.

Keen, M. 2002. *Das Rittertum*. Düsseldorf, Albatros Verlag.

Kitchen, K. A. 1983. *Ramesside Inscriptions, Historical and Biographical V*. Oxford, B.H. Blackwell Ltd.

Kitchen, K. A. 2008. *Ramesside Inscriptions Translated & Annotated: Translations Volume V. Setnakht, Ramesses III. & Contemporaries*. Oxford, Blackwell Publishing Ltd.

Kühnert-Eggebrecht, E. 1969. *Die Axt als Waffe und Werkzeug im alten Ägypten*. Münchner Ägyptologische Studien 15. Berlin, Verlag Bruno Hessling.

Mariette, A. and Maspero, G. 1889. *Monuments Divers acceuillis en Egypte et en Nubie*. Paris, Vieweg

McLeod, W. 1982. *Self Bows and Other Archery Tackle from the Tomb of Tut'ankhamun*. Tut'ankhaum Tomb Series IV, Oxford, Griffith Institute.

Morenz, L. D. 1999. Das Lese-Bild von Königskartusche, Löwe und Stier versus vier Heuschrecken – Königspropaganda und Selbstindoktrinierung der ägyptischen Elite. *Zeitschrift der Ägyptische Sprache und Altertumskunde* 126, 132-140.

Müller, H. W. 1987. *Der Waffenfund von Balâta-Sichem und die Sichelschwerter*. Abhandlungen der Bayerischen Akademie der Wissenschaften, Neue Folgen 97. München, Verlag der Bayerischen Akademie der Wissenschaften.

Petrie, W. M. F. 1901. *Diospolis Parva, the Cemeteries of Abadiyeh and Hu*. London, Special Extra Publication of the Egypt Exploration Society.

Petschel, S. 2004. In S. Petschel and M. von Falk (eds.), *Pharao Siegt Immer. Krieg und Frieden im Alten Ägypten*. various entries. Bönen, Kettler.

Rosselini, I. 1834. *I Monumenti dell'Egitto e della Nubia. Tomo Secondo*. Pisa, Presso Niccoló Capurro & C.

Säve-Söderbergh, T. 1957. *Four Eighteenth Dynasty Tombs*. Private Tombs at Thebes Volume I. Oxford, Oxford University Press.

Schmitz, B. 1977. Bemerkungen zu einigen königlichen Geschenken. *Studien zur Altägyptischen Kultur* 5, 213-219.

Sethe, K. 1903. *Urkunden des Alten Reiches I*. Leipzig, J.C. Hinrich'sche Buchhandlung.

Sethe, K. 1961. *Urkunden der 18. Dynastie. Erster Band. Historisch-biographische Inschriften*. Berlin, Akademie Verlag.

Vila, A. 1970. l'Armement de la Forteresse de Mirgissa-Iken. *Revue d'Égyptologie* 22, 171-199.

von Bissing, F. 1900. *Ein thebanischer Grabfund aus dem Anfang des Neuen Reiches*. Berlin, Verlag Alexander Duncker.

von Deines, H. 1954. 'Das Gold der Tapferkeit', eine militärische Auszeichnung oder eine Belohnung? *Zeitschrift der altägyptischen Kultur und Altertumskunde* 79, 83-86.

Name	Date	Dagger	Axe	Bow	Throwstick?	Khopesh	Mace
Khu-u-Sobek	Sesostris III	X		X			
?	Sobekhotep III		X				
Ameni/ Min-em-hebef-neb	Rahotep			X			
Hor-sekhet	Dedu-mesu		X				
Nehemen	Apophis	X					
Tjuiu (?)	Taa I or II				X		
Ahmose-pen-Nekhbet	Amenhotep I – Tuthmosis II	X	XXX				
Nehmem/Hor-mem	Amenhotep II		X				
Tjenuna (?)	Tuthmosis IV – Amenophis III					X	X
Harsiese	Uasarhata (Libyan Period)		X				

Table 1: Possible examples of gifts of weapons by a king to a private individual.

The *decimatio* in the Roman World

Davide Salvo

Abstract

Decimation was a form of military discipline used by officers in the Roman Army to punish mutinous or cowardly soldiers. A cohort selected for punishment by decimation was divided into groups of ten; each group cast lots and the soldier on whom the lot fell was executed. All soldiers in the selected cohort were eligible for execution regardless of rank or distinction, often by stoning or clubbing in front of other soldiers. The show of violence was the main feature of this practice and everyone had to witness the execution. Those who remained were publicly dishonoured by being given rations of barley instead of wheat and forced to sleep outside the encampment.

The purpose of decimation was to inspire fear and to instil a sense of discipline in the remaining troops. The earliest evidence for this custom is during the war against the Volsci in 492 BC – said to be traditional ('pavtrion') – by both Plutarch and Dionysius of Halicarnassus. It was reportedly revived by Crassus during Spartacus' revolt after long being discontinued. During the late Republic both Octavian and Anthony punished their troops by decimation in Dalmatia and during the conflict against the Parthians respectively. In the imperial age emperor Opilius Macrinus was well-known for centesimatio – *he selected every hundredth man – whilst the emperor Julian, irritated at having been defeated in battle by the Parthians, used decimation to both punish and spur on his soldiers. During the reign of Maximian decimation was also employed to punish Christian soldiers who refused to acknowledge pagan deities as happened to the legio Tebea and its* primicenius Mauritius *(latterly Saint Maurice).*

Introduction

Discipline in the Roman army was extremely rigorous by modern standards, and as Campbell (1984, 300) states: 'traditionally the Romans took great pride in the strict military discipline through which their forefathers had established Roman supremacy'. A Roman general had the power to summarily put to death any soldier under his command. Titus Manlius is reported to have executed his son for attacking the enemy without his permission, declaring 'you have undermined military discipline, by which the Roman state has stood to this day' (Livy, 8. 7.15-16). Polybius observed that punishments and rewards played an important role in Roman military life remarking that: 'considering all this attention given to the matters of punishments and rewards in the army and the importance attached to both, it is no wonder that the wars in which the Romans engage end so successfully and brilliantly' (Polybius, 6.39.11).

The historian of Megalopolis also provides details concerning a variety of military crimes and the way in which they were punished. He makes a distinction between punishments exacted for military crimes (desertion, stealing from the camp, giving false evidence, etc.) and those for unmanly acts (throwing away any of his arms due to fear, leaving his assigned station, etc.). For the former the more common penalties were *fustuarium,* pecuniary fines, demanding sureties and flogging in front of the century (Polybius, 6.37.1-9).[1] For the latter the usual punishment was decimation but also a reduction in rations, eating barley instead of grain (Polybius, 6.38). In this paper we will focus on decimation alone, examining episodes in which it occurred (from the Early Republican age to Late Antiquity) and its wider impact on the behaviour of soldiers.

The Early Republican Age

Dionysius of Halicarnassus (Dionysius of Halicarnassus, *Antiquitates Romanae* 9.50) wrote that in 471 BC the Romans decided to enrol armies and to send out both consuls – Quintius and Appius – against the Aequians and the Volscians respectively. Whilst the army assigned to Quintius carried out his order, the one that went with Appius ignored many of the principles of their ancestors: the soldiers not only played the coward deliberately but also treated their general with contempt. When the time came to engage the army of the Volscians and their commanders had drawn them up in order of battle, they refused to come to grips with the enemy. Instead, both the centurions and the *antesignani* abandoned their posts and fled to the camp, some even throwing away their standards. This behaviour constituted a capital offense according to laws governing the Roman army (others include crimes such as conspiring against a commander, wounding or killing a fellow soldier, entering a Roman camp over the wall, faking illness to avoid battle, espionage, treason and desertion to the enemy in the course of battle).

The punishment which Appius imposed for these acts of insubordination took the form of decimation: Dionysius says that 'the centurions whose centuries had run away and the *antesignani* who had lost their standards were either beheaded with an axe or beaten to death with rods;

---
[1] The punishments were also described in *Digesta* 49.16. See Keppie (1984, 38) and Le Bohec (2008, 78-80).

as for the rank and file one man chosen by lot out every ten was put to death for the rest' adding that decimation was πάτριός ἐστι ('the traditional') punishment among the Romans for those who desert their posts or yield their standards' (Dionysius of Halicarnassus, *Antiquitates Romanae* 9.50.7). Livy, reporting the same episode, wrote that after the defeat of his army Appius resorted instead to beheading the soldiers without their weapons, the *signiferi* who had lost their standards, and the centurions and *duplicari* who fled, adding '*cetera multitude sorte decimus quisque ad supplicium lecti*' (Livy, 2.59).

This is the earliest evidence of this severe custom being imposed, which Polybius describes as follows:

> the tribune assembles the legion, and brings up those guilty of leaving the ranks, reproaches them sharply, and finally chooses by lots sometimes five, sometimes eight, sometimes twenty of the offenders, so adjusting the number thus chosen that they form as near as possible the tenth part of those guilty of cowardice. Those on whom the lot falls are bastinadoed mercilessly...the rest receive rations of barley instead of wheat and are ordered to encamp outside the camp on an unprotected spot. (Polybius, 6.38.2).[2]

All soldiers in the selected cohort were eligible for execution regardless of rank or distinction and were stoned or clubbed to death (*fustuarium*) by their comrades in front of other soldiers. Sometimes the soldiers were beheaded as happened in the case of consul Fabius Rullus who during the Third Samnite War 'chose men by lot and beheaded them in the sight of their comrades' (Frontinus, *Strategemata* 4.35).[3] It should be remembered that even after decimation, the survivors were further humiliated: they were forced to pitch their tents outside the camp and were fed on barley, the food of slaves and animals.[4]

The imposition of capital punishment had to be witnessed by other soldiers as a means of instilling a sense of fear and discipline in the troops as a reference of Menenius, contained in *Digesta*, specifies: '*Qui in acie prior fugam fecit, spectantibus militibus propter exemplum capite puniendus est*' (*Digesta* 49.16.3). The cruel *exemplum* might have a huge psychological impact on the soldiers, and fear for their own lives was supposed to act as a spur encouraging them to fight bravely; for this reason decimation was, according to Polybius, καταπληκτική ('terror striking') (Polybius, 6.38.1).

The show of violence was a characteristic of capital punishments in the Roman world not only in military but also in civil life. Horrific and upsetting death sentences, such as crucifixion for slaves or *damnatio ad bestias*, were carried out in public spaces in order that they were seen by a large number of people, and in some cases into a form of public entertainment.[5] The display of a cruel death was intended as a deterrent to those harbouring thoughts of rebellion, insubordination and anything divergent from the will of the state.

After the war against Volscians, decimation was perhaps employed during the siege of Vei: Livy (5.19.4) simply says that Furius Camillus punished soldiers fleeing in *more militari* but does not mention any other details. The words *more militari* could refer to decimation, as it was considered a custom or *mos*. Frontinus reported another episode: he wrote that an Aquilius[6] '*ternos ex centuriis, quarum statio ab hoste perrupta erat, securi percussit*' (Frontinus, *Strategemata* 4.1.36).

The Late Republican and Early Imperial Ages

After a long period of hiatus, decimation was again employed by Crassus in the Slave War of Spartacus: the commander is described by the ancient sources as having revived the ancient punishment which had long fallen into disuse (Plutarch, *Crassus* 10.4-5). Our sources for this incident are Appian and Plutarch, but their tales diverge somewhat from each other. On the one hand Appian writes that Crassus punished with decimation the two legions of consuls defeated before his coming. The writer also adds that some historians had claimed that the commander punished eight legions disbanded by rebels, but this version seems improbable,[7] as it alleges that Crassus ordered the execution of more than 4000 soldiers! It may possibly originate from a source hostile to the military leader. On the other hand, Plutarch reports that Crassus ordered his lieutenant Mummius not to fight; but that this order was disobeyed and Mummius was defeated by slaves. Furious at this act of insubordination, Crassus chose 500 soldiers from amongst those who were first to flee and decimated them for their cowardice. From a fragment of *Historiae* of Sallust it is known that soldiers were clubbed ('*necat fusti*') (Sallust, *Historiae* 4.Fragment 22).

This act of Crassus had a particular significance. He would revert to the *mos maiorum*, this punishment being a custom attributed to their ancestors. He may have been

---

[2] During the military campaign in Armenia, Corbulo punished one detachment for its disobedience by ordering them to encamp outside the rampart but he didn't order *decimatio* (Tacitus, *Annals* 13.36.3 and Frontinus, *Stratagems* 4.1.21).

[3] It could be possible (although it would be very difficult to prove) that the 50+ decapitated skeletons, all males, found in 2004-2005 at 1-3 and 6 Driffield terrace in York (see the 34th annual report of York Archaeological Trust at website www.orkarchaeology.co.uk/about/annrep.), who had probably been executed, were victims of *decimatio*. I wish to thank Dr P. Ottaway who kindly suggested to me that this unusual cemetery area was for soldiers who had disobeyed in some way.

[4] For barley instead of wheat see Plutarch, *Antony* 25; Frontinus, *Stratagems* 4.25.

[5] I analyzed this aspect – referred to as the Slave Wars – in Salvo (2007).

[6] The identity of that Aquillius is not certain. Klebs (1896) avoided determining his identity, but Morgan (2003, 499 n. 37) deemed it likely to be the identification of C. Aquilius Florus (consul in 259 BC), M. Aquilius (consul in 129 BC) or M. Aquilius (consul in 101 BC).

[7] Bertinelli (1989, 356) writes that 'Il sacrificio esemplare di un numero così elevato di soldati (quasi un'intera legione), in una situazione di grave pericolo pubblico si rivela assurdo'. See also Salvo (2007, 98).

modelling himself on his grandfather, Marcus Licinius Crassus, nicknamed *Agelastos* ('serious') for his austerity and strict manners. At this point Crassus was building up his political career and needed the support of the Senate: for this purpose he established this reputation as a severe and righteous man, the defender of traditional culture as threatened by *nefarii* ('non-Roman slaves'). It is significant that the following year – 70 BC – he became consul with Pompey. The decimation, said to be τι πάτριον ('something traditional') by Plutarch, played an important role not only in strengthening this image but above all in spurring on the soldiers. It seems to have been successful in that, Appian wrote that after the *decimatio* had been carried out, the soldiers feared Crassus more than the enemy and won an important battle against the slaves (Appian, *Bella Civilia* 1.119).

During the Civil Wars *decimatio* became common once more, and was retained under the Empire with many commanders and Emperors using it as punishment. In 49 BC Caesar punished (Cassius Dio, 41.35.5; Appian, *Bella Civilia* 2.7) the Ninth legion with a capital sentence because of its mutiny at Piacenza. Amongst all sources referring this episode (Plutarch, *Caesar* 17; Cassius Dio, 41.35.5; Suetonius, *De Vita Caesarum* 69; Lucan, *Pharsalia* 237-373; Appian, *Bella Civilia* 2.47), only Appian makes it clear that the punishment was decimation, saying that after containing the mutiny, Caesar let the rebels choose 120 soldiers to be decimated.

In 39 BC Calvinus lead a military campaign in Spain against Cerretani. He won many victories but was defeated in a battle because his lieutenant was killed after being abandoned by his soldiers. For this reason he decimated two centuries by punishing many centurions, amongst which can be identified the chief centurion ('*primipilus*') Vibilius (Cassius Dio, 48.42.1-3; Velleius Paterculus, 2.78.3). Cassius Dio said that after this punishment Calvinus 'obtained the first place concerning military discipline after Crassus' (Cassius Dio, 48.42.3) and Velleius Paterculus (Velleius Paterculus, 2.78.3) wrote that 'Calvinus Domitius....executed a rigorous act of discipline comparable with the severity of the older days'. As with Dionysius and Plutarch, Velleius puts in evidence that decimation was considered an act of 'the older days' and perhaps it was used by Calvinus in order to establish a reputation of traditional man reverting to the past in the same way as Octavianus, of whom Calvinus was a supporter (Cassius Dio, 48.42.4).

In fact when Octavianus went to Dalmatia in order to suppress a revolt of *Dalmatii* in 35 BC he had great difficulty in overcoming the rebels and disciplining his troops. For this reason he decimated groups of soldiers who deserted and gave out barley to others (Cassius Dio, 49.38.4). Suetonius reports the same episode, writing that the Emperor disciplined the army strictly, reverting to ancient customs ('*ad antiquum morem*') and in doing so he disbanded the undisciplined Tenth legion, decimated its soldiers and forced the defeated cohorts to eat barley rations in order to punish both the centurions and the soldiers who had fled (Suetonius, *Divus Augustus* 24).[8] Octavianus, like Crassus, showed this austerity in military matters in order to build up his image of *optimus princeps*, reverent to *mos maiorum*, and to intimidate his opponent Anthony.

In the 1st century AD decimation was also used during the war against Tacfarinas under the reign of Tiberius. In AD 17 Tacfarinas, a Numidian soldier, rose up in revolt against Rome. Furius Camillus defeated him but in the following year the rebel again defied Rome: he besieged a Roman garrison of which Decrius was in command. The soldiers of the cohort resident in the fort fled and left Decrius to be killed by the enemy. Lucius Apronius, the new governor of Africa, angered by the cowardice of the fugitives, decimated the cohort. Tacitus subsequently wrote that 'he restored an old custom that became uncommon in that time' (Tacitus, *Annales* 3.21).[9] It is also reported that Caligula sought to decimate the legions that had rebelled against his father Germanicus when he was a child (Suetonius, *Gaius Caligula* 48).[10]

After the death of Nero and the subsequent entry of Galba into Rome a cruel episode occurred (Plutarch, *Galba* 15.5-9; Suetonius, *Galba* 12.2; Cassius Dio, 64.3.2; Tacitus, *Historiae* 1.6.2; 31.2; 37.2-4; 51.5; 87.1). Nero had recruited a legion – the *I Classicorum Adiutrix* – from units drawn from the navy. When Galba become Emperor, he decided to disband the legion and to allow the soldiers to go back into their previous roles. However, the soldiers, who ran forward to meet him when he entered the city, refused to do so, and for this reason Galba ordered the entire legion to be decimated.[11] This episode is very obscure. Morgan (2003) supports the theory that the decimation of *legio classiaria* was necessary for Galba's acknowledgement as *iusta legio*. In this way its loyalty and discipline might have been guaranteed and it could be left to defend the city. In the 3rd century AD, the Emperor Opilius Macrinus usually decimated the soldiers, but sometimes he punished only the hundredth man, and for this he coined a new word: *centesimatio*. He also said that he was merciful in putting to death only one in one hundred instead of one in ten or one in twenty ('*vicesimatio*') (Scriptores Historiae Augustae, *Macrinus* 12).[12]

Decimatio in the East

In the East, *decimatio* was used twice during the many wars against the Parthian Arsacids and Sassanids from the 1st century BC onwards. After the defeat of Carrhae, the Parthians became a thorn in the side of Rome,

---

[8] See Le Bohec (2008, 78).
[9] See Salvo (2007, 98-99).
[10] Burnett (1989) and Hurley (1993) dismiss this episode out of hand.
[11] Amongst the four sources only Suetonius and Cassius Dio reported as a fact the *decimatio*. See Morgan (2003, 498).
[12] The cruelty of Macrinus and his son Diadumenos is attested by Scriptores Historiae Augustae, *Diadumenus Antoninus* 8.2-3 and Aurelius Victor, *Epitome de Caesaribus* 22.

inflicting frightful defeats.[13] Although Caesar was planning a great military expedition against Parthia prior to his murder in the 1st century BC (Plutarch, *Caesar* 58.6; Cassius Dio, 43.51.1; Suetonius, *De Vita Caesarum* 44 and 49; *Divus Augustus* 8; Appian, *Bella Civilia* 3, 23-24, 26, 92 and 144), his plan was taken up by Anthony. In 36 BC Anthony went with a huge army to Armenia where he was joined by the troops of various allies in order to make war on the King of Parthia. He left Armenia and went straight on to Media, where he besieged the city of Praaspe. It was a difficult operation to undertake as his war machines had been destroyed by the enemy. In spite of being besieged, the Parthians attacked the Roman camp outside the city walls and terrified the soldiers.

Furious at the cowardice of his soldiers, Anthony decided to decimate them. Plutarch writes that 'he divided the bulk of soldiers into groups of ten and made to kill one for each group selected by lot. The remaining received rations of barley instead of wheat' (Plutarch, *Antonius* 39.9). The same story can be found in Cassius Dio (Cassius Dio, 49.27.1), whilst Frontinus gives us more details, writing that:

> Marcus Antonius, when fire had been set to his line of works by the enemy, decimated the soldiers of two cohorts of those who were on the works, and punished the centurions of each cohort. Beside this, he dismissed the commanding officer in disgrace and ordered the rest of the legion to be put on barley rations. (Frontinus, *Strategemata* 4.37).

In AD 224 the reign of Arsacids came to an end and Ardashir established the Sassanid dynasty, the new enemy of Rome along the Eastern *limes*.[14] In the 4th century AD the Emperor Julian the Apostate made another great expedition against the Sassanid Empire.[15] The Emperor crossed the Euphrates river and conquered many places in Assyria, but while he was besieging the city of Pirisabora, the Sassanid commander Surena attacked his front line, dispersing the soldiers and taking a standard. The Emperor was irate at this perceived affront, demoted ten soldiers amongst those who fled, and 'inflicted on them a capital sentence in accordance with ancient laws' (Ammianus Marcellinus, 24.3.2). It is likely that Ammianus is referring to decimation, the only kind of military punishment considered traditional. We have already seen how crimes similar to those perpetrated by soldiers of Julian were punished with decimation by other commanders such as Appius Claudius and Anthony.

In AD 70, during the siege of Jerusalem, it seems that a legion was almost decimated after a Jewish ruse lead to a Roman defeat (Josephus, *Bellum Judaicum* 5.109-119). The soldiers of *legio VI Ferrata* failed to carry out the orders of Titus and they were attacked by the Jews. The Emperor was furious and threatened to punish the soldiers for their reckless disobedience in the traditional manner. However the legions interceded on behalf of their fellow soldiers (Josephus, *Bellum Judaicum* 5.125-126) and Titus relented, realizing that punishing numerous soldiers had potentially serious consequences (Josephus, *Bellum Judaicum* 5.128-129).[16] I agree with Morgan's (2003, 500 n.43) statement that 'Titus may have contemplated decimating *VI Ferrata* or some part of it for reckless disobedience during the siege of Jerusalem'. There are indications that Titus wanted to carry out a decimation: in his speech to his generals he is reported to have said that 'laws invariably punish with death the very slighted breach of discipline, whereas now they have beheld a whole corps quit the ranks' (Josephus *Bellum Judaicum* 5.124). It is also clear from the various Republican examples described above that the traditional punishment for breaking the military rules was *decimatio*. Moreover, the fact that legions interceded on behalf of their comrades appears to suggest that they were equally terrified at the thought of having to either club their fellow soldiers to death or witness such a macabre sight.

Some of these examples suggest that from the Late Republican age decimation had two motives. On the one hand it functioned as an effective means of disciplining the troops. Its significance was also political, however, in that its application was useful for improving the reputation of commanders emphasizing their devotion to tradition and the *mos maiorum*.

Saint Maurice and the *legio Tebea*

In Late Antiquity, decimation could also have a religious implication (as in the story of Saint Maurice): it was considered a martyrdom, the instrument proving the Christian faith, and as such possessed a different meaning from Roman tradition. The martyrdom of Mauritius and the *legio Tebea* was the subject of *Passio Acaunensium martyrum*, of which two versions survive, one written by Eucherius of Lugdunum[17] in the 5th century AD[18] and the other by an anonymous author, the chronology of whom is controversial.[19] The authenticity of *Passio* is a matter

---

[13] In the time of Nero the defeat of Cesennius Petus resulted in Roman legions passing under the Parthian yoke. See Suetonius, *Nero* 39.1; Eutropius, 7.14; Rufus Festus, 20 and Orosius, 7.7.12. Tacitus, *Annales* 15.15 wrote that the passage under the yoke was a *rumor*. See also Scardigli (2007, 157-159).
[14] In the 3rd and 4th centuries AD Romans and Sassanids clashed on many occasions. One Roman Emperor, Valerianus, was even forced to work as a slave after being defeated by Sassanid Emperor Shapur I. See Scriptores Historiae Augustae, *Valerianus* 2.1 and 4.1 and Gallienus, 1.1 and 21.5; Eutropius, 9.7; Aurelius Victor, *Epitome de Caesaribus* 32.5-6; Lactantius, *De mortibus persecutorum* 5.2-6; Orosius, 7.22.4; Zosimus, 1.36.2-7.
[15] For the eastern military campaign of Julian see Le Bohec (2006, 72-76).

[16] On the severity of Titus in maintaining the military discipline see Campbell (1984, 305) who wrote that 'few Emperors were as strict as Titus'.
[17] The text was edited by Krusch (1896).
[18] See Chevalley (2005, 424): 'La datation de la Passion d'Eucher de Lyon ne pose guère de difficultés, puisqu'elle est solidement située entre 443 et 451'.
[19] The text was edited by Chevalley (1990). The date of composition is very controversial and the proposed chronologies variously dated the anonymous *Passio* between the 6th century AD and the Carolingian Renaissance, whilst Chevalley (2005, 423) hypothesized that: 'la

of debate, however, since some scholars argue that the history is a forgery, an oral legend created by Theodore the Bishop of Octodurus, put in writing by Eucherius[20] and the anonymous author.[21]

It is worth remarking however that even if the *Passio* is a fake,[22] the fact that *decimation* features in the narrative relating to the *legio Tebea* suggests that it was still considered a traditional punishment in the 3rd/4th centuries AD and might still be applied. Perhaps in the minds of those who formulated this story, this sort of punishment was the traditional (and pagan) way of ordering authority, displaying the cruelty of *impius* Emperors in contrast to the courage of the Martyrs.

Both versions of *Passio* state that the *legio Tebea*, made up of Christians, went to the East to help Maximianus. When the soldiers of this legion were at Acaunum on the Alpes,[23] they twice refused to obey the Emperor. Once Maximianus had received these denials, he ordered that the legion be decimated. Mauritius, the *primicenius* of this legion, played an important role encouraging his comrades to refuse the order and face death. The two versions of *Passio* give different reasons for this punishment: according to Eucherius, the soldiers of this legion refused to kill other Christians, whilst the anonymous author says that Maximianus' anger was provoked by the fact that they had refused to acknowledge pagan deities. Another difference between the two texts concerns the date of punishment: for the anonymous writer it was around AD 285-286, during the revolt of Bagaudes[24] in Gaul, whilst for Eucherius it occurred during the persecution of Diocletian and Maximianus (AD 303-305).[25]

Conclusion

This seems to be the last evidence of this practice, which was abandoned once Christianity became the main religion of Rome. From its original role as a form of military punishment the decimation gradually acquired new connotations: an affirmation of traditional values during the Late Republican age and Early Empire, and *instrumentum dei* through which the Christians confirmed their faith and loyalty to God in the episode of *legio Tebea*.

Bibliography

Antonini, A. 2005. Les origins du monastère de Saint-Maurice d'Agaune- un héritage à étudier et protéger. In O. Wermelinger, P. Bruggisser, B. Näf and J. -M. Roessli (eds.), *Mauritius und die Thebäische legion. Akten des internationalen Kolloquiums Freiburg, Saint-Maurice, Martigny, 17-20 September 2003*, 331-342. Fribourg, Academic Press.

Bertinelli, M. G. 1989. Commento alla Vita di Crasso. In M. G. Bertinelli, C. Carena, M. Manfredini and L. Piccirilli (eds.), *Plutarco. Vite di Nicia e Crasso*, 317-430. Milano, Mondadori.

Burnett, A. 1989. *Caligula: the Corruption of Power*. London, B.T. Batsford.

Campbell, J. B. 1984. *The Emperor and the Roman Army. 31 BC-AD 235*. Oxford, Clarendon Press.

Carrié, J. -M. 1993. Eserciti e strategie. In A. Momigliano and A. Schiavone (eds.), *Storia di Roma*, vol. III (1), 83-154. Torino, Einaudi.

Carrié, J. -M. 2005. Des Thébains en Occident? Histoire militaire et hagiographie. In O. Wermelinger, P. Bruggisser, B. Näf and J. -M. Roessli (eds.), *Mauritius und die Thebäische legion. Akten des internationalen Kolloquiums Freiburg, Saint-Maurice, Martigny,17-20 September 2003*, 9-35. Fribourg, Academic Press.

Chevalley, È. 1990. La Passion anonyme de saint Maurice d' Agaune. Èdition critique. *Vallesia* 45, 37-120.

Chevalley, È., Favrod, J. and Ripart, L. 2005. Eucher et l'Anonyme: les deux Passions de saint Maurice. In O. Wermelinger, P. Bruggisser, B. Näf and J. -M. Roessli (eds.), *Mauritius und die Thebäische legion. Akten des internationalen Kolloquiums Freiburg, Saint-Maurice,*

---

passion anonyme est antérieure à celle qu'Euchere rédigea dans les années 440'.

[20] See Van Berchem (1956, 26). This scholar calls the authenticity of account as found in the *Passio* in question as the tale seems to have many inconsistencies such as the punishment by *decimatio,* that for him 'elle (*scil.* decimation) n'était plus concevable à la fin du IIIe siècle, où, dans une société profondément évoluée, la discipline militaire avait elle-même changé'. Carriè (2005, 14) also considers decimation of *legio tebea* an 'anachronisme...tombée en désuétude depuis des siècles'. Van Berchem also dismisses out both the reference of *Historia Augusta* concerning Macrinus and the subsequent reference of Ammianus. The author strongly disagrees with him because there are not any reasons for dismissing out of hand the truthfulness of *Historia Augusta*'s reference and we believe that Ammianus' reference concerns the *decimatio* (see above).

[21] See Van Berchem (1956), Carriè (2005, 13-14) and mainly Roessli (2005), who goes back to the 16th century AD in order to re-enact the debate on the authenticity of Passion of *Acaunum*. These scholars put in evidence the fact that the authenticity of the account of Maurice was never called into question until the age of Reform, when the protestant Flacius Illyricus, in his *Centuries of Magdebourg*, considered the story of Maurice a forgery, touching off the response of Catholics.

[22] Carriè (1993, 143-144) writes that 'la credibilità dell'episodio, che pure conta ancora alcuni accaniti difensori, sopravvive difficilmente allo studio in cui Denis Van Berchem ha ricostruito la storia e le motivazioni della pia finzione'. The same opinion appears in Woods (1994) *contra* Curti (1959, 311) who writes that 'la storicità del martirio di Maurizio e dei suoi compagni non può essere messa in dubbio, anche se le vicende quali sono presentate da Eucherio non siano immuni da amplificazioni'. Dupraz (1961), Girgis (1985) and Le Bohec (2006, 285) also recognized the authenticity of the episode of *legio Tebea*

[23] *Agaunum* is the modern town of Saint Maurice in Canton Valais, Switzerland where, for honouring the martyrdom of *legio tebea*, the 'Abbaye de Saint-Maurice d'Agaune' was built by Sigismond, King of Burgundians in AD 515 (but in the 4th century AD the Bishop of Octodurus (Martigny) had already built a sanctuary). In medieval times the abbey increased in popularity. See Theurillat (1954) and Antonini (2005).

[24] For this revolt see Eutropius, 9.20; Aurelius Victor, *Epitome de Caesaribus* 39.17; Ammianus Marcellinus, 28.2.10; Orosius, 7.25.2; Zosimus, 6.2.5; Zonaras, 12.31.

[25] See Carrié (2005, 10-11).

Martigny, 17-20 September 2003, 423-438. Fribourg, Academic Press.

Curti, C. 1959. La *Passio Acaunensium martyrum* di Eucherio di Lione. In *Convivium Dominicum: studi sull'eucarestia nei Padri della Chiesa antica e miscellanea patristica*, 298-325. Catania, Centro di Studi sull'antico Cristianesimo.

Dupraz, L. 1961. *Les Passions de Maurice d'Agaune. Essai sur l'historicité de la tradition et contribution à l'étude de l'armée pré-dioclétienne et des canonisations tardives de la fin du IVe siècle.* Fribourg, Studia Friburgensia, n.s. 27.

Girgis S. 1985. *The Theban Legion in Switzerland.* Zürich-Wien, St Pachon's Publications.

Hurley, D. 1993. *An historical and historiographical Commentary on Suetonius' Life of C. Caligula.* Atlanta, Scholars Press.

Keppie, L. 1984. *The Making of Roman Army.* London, B.T. Batsford.

Klebs, E. 1896. s.v. Aquilius. *Pauly-Wissowa Realencyclopädie der Classischen Altertumswissenschaft* 2, 3-322.

Krusch, B. 1896. *Monumenta Germaniae Historica, Scriptores Rerum Merovingiarum* (MGH SS rer Mer), vol. III, 32-41. Hannover.

Le Bohec, Y. 2006. *L'armée romaine sous le Bas-Empire.* Paris, Picard.

Le Bohec, Y. 2008. *L'esercito romano. Le armi imperiali da Augusto alla fine del terzo secolo.* 8th Italian edition. Roma, Carocci.

Morgan, M. G. 2003. Galba, the Massacre of the Marines and the Formation of Legion I Adiutrix. *Athenaeum* 91, 489-515.

Roessli, J. -M. 2005. Le martyre de la Légion Thébaine et la controverse autour de l'historicité du XVI eau XVIIIe siècles. In O. Wermelinger, P. Bruggisser, B. Näf and J. -M. Roessli (eds.), *Mauritius und die Thebäische legion. Akten des internationalen Kolloquiums Freiburg, Saint-Maurice, Martigny, 17-20 September 2003*, 193-210. Fribourg, Academic Press.

Salvo, D. 2007. Rivolte servili e spettacolarizzazione della violenza. *Hormos* 8, 93-102.

Scardigli, B. 2007. Corbulone e dintorni (Tac., Ann. XV 15). In M. A. Giua (ed.), *Ripensando Tacito (e Ronald Syme). Storia e storiografia, Atti del Convegno Internazionale (Firenze, 30 novembre-1dicembre 2006)*, 153-160. Pisa, ETS.

Theurillat, J. M. 1954. L'Abbaye de St-Maurice d'Agaune des origines à la réforme canoniale (515-830). *Vallesia* 9, 1-128.

Van Berchem, D. 1956. *Le martyre de la legion thébaine. Essay sur la formation d'une legend.* Basel, Friedrich Reinhardt.

Woods, D. 1994. The Origin of the Legend of Maurice and the Theban Legion. *Journal of Ecclesiastical History* 45, 385-395.

The Development of Warfare and Society in 'Mycenaean' Greece

Stephen O'Brien

Abstract

Warfare has been of concern to Aegean prehistorians since the birth of the discipline, yet few studies have attempted to place warfare in its full social context. For decades the societies of the region have been described as 'chiefdoms' or 'states' in accordance with neoevolutionary typologies. More recently, scholars of the Late Bronze Age in Crete have challenged the neoevolutionary approach and suggested alternative forms of social organisation, yet these concepts have so far had comparatively little impact upon the archaeology of the Greek mainland in the same period. This paper will attempt to tackle both problems by adopting a dual approach. Alternative approaches to the social complexity of 'Mycenaean' societies from c.1700-1100 BC will be outlined, and these will be used to assess the role which violence, and the social use of violent concepts, played in the construction and maintenance of social relations. It is hoped that this will both reintegrate the study of warfare and violence into the broader research questions of Aegean prehistory, and contribute towards the ongoing debate on archaeological approaches to social complexity.

Introduction

Warfare has been a subject of study in Aegean archaeology since the very beginnings of the discipline, with the original excavations performed by Heinrich Schliemann at Troy, Mycenae, and at Mount Aëtos on Ithaca being part of his project to prove the literal historicity of the Trojan War as described in Homeric Epic (Schliemann 1878). When Arthur Evans discussed the 'palatial'[1] societies of 'Minoan'[2] Crete, he did so with reference to Thucydides' (1.4) thalassocracy of King Minos, imagining a peaceful island culture (Evans 1928, 571) which simultaneously exercised a dominant role in the region surrounding the island (Evans 1935, 887). More recent generations of scholars have abandoned the concept of an historic Trojan War and that of a peaceful but thalassocratic 'Minoan Crete', yet warfare has remained a prominent feature in studies of later Aegean prehistory.[3]

While warfare has been seen as an implicit feature of Mycenaean society, there has been comparatively little work undertaken by Aegean prehistorians which has examined the relationships between warfare and society, with warfare not featuring at all in one recent survey of research trends in the discipline (Tartaron 2008). For much of the 20th century, research into Mycenaean warfare was dominated by a concern for the creation of typological catalogues of military equipment rather than investigating the social and cultural aspects of inter-human violence in the period (Sandars 1961; Buchholz 1962; Sandars 1963; Snodgrass 1964; Buchholz and Wiesner 1977; Buchholz 1980; Höckmann 1980; Avila 1983; Fortenberry 1990; Kilian-Dirlmeier 1993; Papadopoulos 1998; Hope Simpson and Hagel 2006).

Due in part to a greater concern with warfare in both archaeology and anthropology as a result of political events during the last two decades (Vandkilde 2003), a more socially-situated archaeology of Mycenaean warfare has recently begun to emerge, as demonstrated by contributions in the proceedings of the 'POLEMOS' conference (Laffineur 1999), and the work of Kate Harrell (2009), Angelos Papadopoulos (2006), and Barry Molloy (2006; 2008; 2009; 2010).

Over a similar period, the ways in which the societies of the Aegean Late Bronze Age are envisioned have also changed. Influenced by the neoevolutionist typologies of band > tribe > chiefdom > state (Sahlins and Service 1960, 37) and egalitarian > ranked > stratified > state (Fried 1967), Aegean prehistorians interpreted the appearance of 'palatial' structures in the region as signifying the development of 'states' (Renfrew 1972 [2011]; Cherry 1986).[4] Widespread disillusionment with the neoevolutionary paradigm has more recently led to a number of alternative versions of Late Bronze Age Aegean societies being put forward, particularly in Cretan archaeology, which has seen the development of the concepts of the decentralised state (Knappett 1999), heterarchical organization (Schoep 2002), factionalism (Hamilakis 2002; Schoep 2002; Wright 2004), a more dynamic model of hierarchical organisation (Adams 2004), and an application of the 'dual-processual' model developed for Mesoamerican societies (Blanton et al.

---

[1] The term 'palace' has been recognised as inadequate when applied to Aegean prehistory, and a number of alternatives proposed (Hitchcock 2000, 46-47), although none has attained general currency.

[2] 'Minoan' and 'Mycenaean' are terms which tend to homogenise very real differences in culture and divergent historical trajectories visible within the areas they claim to delineate, and their use here should not be seen as ignoring the diversity of societies found within the Late Bronze Age Aegean.

[3] See e.g. Robert Drews (1993) and Jan Driessen and Colin Macdonald (1997), along with their critics (Littauer and Crouwel 1996; Dickinson 1999; Warren 2001).

[4] Colin Renfrew (1975) later used the concept of the 'Early State Module', which represents a stage in-between those of the 'chiefdom' and 'state' in a neoevolutionary sequence.

1996) to the Aegean (Parkinson and Galaty 2007). To date such work has left the societies of the Greek mainland largely untouched.[5] However, with the concepts of 'chiefdom' and 'state' being used relatively uncritically (e.g. Dickinson 1994; Wright 1995; Shelmerdine 2001; Shelmerdine and Bennet 2008; Wright 2008; Nakassis, Galaty and Parkinson 2010), a reassessment of the nature of the societies of the Greek mainland in this period cannot help but have an impact upon our interpretations of the role of violence in those societies.

Evolutionary approaches to matters of 'social complexity' have also characterised studies of warfare and its relationship to human societies. The influential work of Harry Turney-High (1949) claimed a distinction between 'primitive' and 'civilised' warfare, separated by a 'military horizon' of social organisation at which a warfare of pitched battles, military formations, and command by officers begins to take place. The theory explicitly suggests military organisation as a driver of the development of social hierarchy (Turney-High 1949, 252-253). The social organisation which brings about the 'military horizon' is equated with the state (Turney-High 1949, 231). A strong association between the practice of warfare and state formation has subsequently been a prominent feature of archaeological and anthropological literature (e.g. Fried 1961; 1967; Carneiro 1970; Flannery 1972; Service 1975; Webb 1975; Webster 1975; Claessen and Skalník 1978; Haas 1982; Cohen 1984; Otterbein 1989; Earle 1997; Gat 2006). Even the briefest survey of this literature will, however, show that the relationship between warfare and state formation is not held to be a straightforward one (Cohen 1984, 332; Otterbein 2004, 11-12; Allen and Arkush 2006; Underhill 2006, 3). The consequences of non-evolutionary models of social change on the theoretical relationship between warfare and 'social complexity' have yet to be explored in detail: the recent study by Azar Gat (2006) remains structured by the terminology of 'tribes', 'chiefdoms' and 'states'.

The purpose of this paper is to draw together these diverse strands of research. The possibility of forms of social organisation existing in the Late Bronze Age Greek mainland which are not explicable by a neoevolutionary typology of 'chiefdoms' and 'states', will be outlined. The evidence of warfare and other practices of inter-human violence will be presented alongside this, in the hope of showing what significance, if any, such practices had upon social change. In its analysis, this paper will serve as an initial effort at moving the study of warfare and violence away from an insular concern with weaponry and tactics (as per Parker Pearson 2005, 21) and realigning it with the broader research questions of Aegean prehistory.

Early Mycenaean Period: Middle Helladic III-Late Helladic IIB (c.1750-1400 BC)

The traditional view of the Middle Helladic period, from which Mycenaean culture emerged in MH III/LH I, is one of 'simple' societies which lack evidence for neoevolutionary criteria of 'complexity': small, village-sized settlements, a lack of large or important buildings, and small-scale industries producing a limited range of objects (Dickinson 1977, 32-34). This is considered to represent a lower 'cultural level' than other parts of the region, leading to the epithet 'The "Third World" of the Aegean' (Dickinson 1989, 133). The one exception is the site of Kolonna on Aegina in the Saronic Gulf, which possesses features including a building complex of monumental scale used from MH I/II-LH I/II (Gauß, Lindblom and Smetana 2011), fortification walls, significant quantities of imported Cycladic and Minoan pottery, a 'Shaft Grave', and an ashlar block with a Cretan-type 'mason's mark', that have led to Kolonna being categorised as a more 'complex' society than that of much of the MH mainland, even being described as a 'state' (Niemeier 1995). This picture may be substantially revised, however, if a different views of 'social complexity' and how it may be manifest materially is taken (Georgousopoulou 2004; Wolpert 2004). A reconsideration of the nature of societies in the MH period of the Greek mainland may yield significant insights into the emergence of the Mycenaean culture at the end of that period.

The occurrence of new social and cultural elements over a wide geographical area in the MH III/LH I period marks the start of both Mycenaean culture and the transition to the Late Bronze Age of that region. These social and cultural elements include the arrival of new mortuary practices ('Shaft Graves', tholos tombs, and chamber tombs), and the construction of larger buildings of diverse form and poorly-understood function at Ano Englianos (Pylos) (Nelson 2001, 194-200), Menelaion (Catling 2009, 23-32), Tiryns (Kilian 1987), Eleusis, Kakovatos, and Thermon (Wright 2008, 249).[6] These developments have been used to suggest a move upwards in the 'level' of social complexity, as signifiers of greater centralisation and the emergence of mainland 'chiefdoms' (Mee and Cavanagh 1984; Wright 1995). The assumption of chiefdoms is not necessary, however, and James Wright (2008, 244) mentions in passing the possibility of more 'corporate' forms of organisation. Viewing the emergence of Mycenaean culture as being less an *increase* in the level of social complexity as a change in the *kind* of social complexity present allows for the potential diversity of forms of social organisation present in the Greek mainland at this time, and may account better for the nature of the archaeological record.

---

[5] Notable exceptions are David Small (1998; 1999), William Parkinson and Michael Galaty (2007) and Robert Schon (2010), which make detailed analyses of the societies of the Late Bronze Age Greek mainland.

[6] Rodney Fitzsimmons (2011, 102-103) correctly identifies the factoid of an early building at Mycenae as emanating from archaeologist's beliefs that such a structure *ought* to be there, rather than from anything present in the archaeological record.

Amongst the new cultural elements found in the early Mycenaean period are a number of items of military equipment, including the Type A and B swords (Sandars 1961; Kilian-Dirlmeier 1993), boar's tusk helmets (Borchhardt, J. 1977) and iconographic evidence for the use of chariots (Crouwel 1981, 59-61) and shields (Borchhardt, H. 1977). These objects have often led to the conclusion that there was a connection between the emergence of Mycenaean societies and the practice of warfare or other types of inter-human violence. Violence was not a phenomenon which arrived in the Greek mainland during the early Mycenaean period, as earlier archaeological indications of its practice have previously been observed (Hood 1954, 11; Nordquist 1987, 21; Arnott 1999, 499-500). A number of researchers have, on the basis of ethnographic as well as archaeological evidence, asserted that high levels of inter-human violence were the norm for most societies in the past (Keeley 1996; LeBlanc 2003; Gat 2006; Pinker 2012). This, however, appears to rely on a methodologically unsound uniformitarian principle (non-state societies in the contemporary world are not in the same context as past societies), and underplays the existence of contemporary non-state societies with low homicide rates (Dentan 1988; Ferguson 1997; Kelly 2000). The archaeological record is a poor resource for assessing the level of violence in a past society, due to the problems of evidence discussed by Slavomil Vencl (1984) and Anthony Harding (2007, 31-40), and claims of high levels of violence based on such evidence have been found to be over-stated (Carman and Carman 2005, 219). There appears to be no obvious case for assuming that human violence was less complex or variable in the past than it is in the contemporary world (Kelly 2000; Otterbein 2004; Thorpe 2005). Such variability makes claims of endemic levels of human violence due to the operation of 'human nature' dubious, given that 'human nature' does not – in this case at least – appear to be a historical constant.

Under these circumstances, what can be said about the significance of the military items and violent iconography which are present in several areas of the Greek mainland at the start of the Mycenaean period? While the changes to mortuary practice may be interpreted as representing the emergence of centralised ruling families (Mee and Cavanagh 1984; Kilian 1988; Graziadio 1991) they are increasingly seen as a social strategy aimed at creating differentiation on the part of an elite which had not constructed relationships of centralised power (Voutsaki 1995). The presence of a number of items of Cretan manufacture or inspiration in mainland graves – including the bronze swords first seen at Malia on Crete in MM II (Kilian-Dirlmeier 1993, 14-17) – has led to Crete being ascribed a significant role in the changes which occurred on the mainland (Sherratt and Sherratt 1991). Molloy (this volume) mirrors this view in the military sphere, suggesting that the 'martial' character of the mainland at this time was a minor regional variation of events on Crete, with the commonality of material between Crete and the mainland representing a commonality of 'military systems' and the practice of violence.

It is possible, however, that the role of Crete in mainland developments may have been given too central a position. Aaron Wolpert (2004, 131) argues that the material expression of Cretan societies did not possess some irrepressible power which reconfigured the Greek mainland, while Sofia Voutsaki (1999, 113) notes that Cretan items in mainland graves were redefined and incorporated into mainland norms. We may also critique the idea that a commonality of military items between two regions necessarily means a commonality of 'military systems': the chariot was put to a very different military use in the Aegean compared to western Asia and north Africa, despite commonalities of design (Littauer and Crouwel 1983; 1996).[7] While there can be little doubt that the introduction of new military items from Crete was significant, we must be cognisant that the socio-cultural environment into which these items came may have had a profound impact upon their use.

The new depositional practices of the early Mycenaean mortuary sphere may be interpreted as an attempt to create a new set of values for the societies in question, with attendant effects upon power (Damilati and Vavouranakis 2011, 38-39). The deposition of swords in the graves of the period, becoming more 'hyperbolic' as time progressed, is not an act which reflects some claim to power or status, but is, rather, an act which creates the sword as an object of value and gives significance to its other uses and meanings. Depositions of swords and other military items in graves are rare – although not unknown (Dimopoulou 1999) – on Crete, where the deposition of swords in sanctuaries and shrines was a more usual practice (Molloy this volume). Such divergent depositional practices may indicate different values and social significances being ascribed to swords in the societies of different parts of the Aegean.

The iconographic art of the early Mycenaean period sees the appearance of gold rings and seals from the Argolid and Messenia which depict individualised or small-group combats with swords.[8] Similar scenes are found on Crete from approximately the same period,[9] suggesting that Crete may have been the origin of this motif. Such depictions need not be read as accurate representations of combat practice, although the presence of other kinds of violent individualised confrontation (e.g. the Akrotiri and Tylissos 'boxer frescoes', the Ayia Triada 'boxer vase') may lend some credence to the idea of 'duels' being fought with swords. The swords of this period, in addition to being suitable for larger confrontations, would also have performed well in smaller or individualised combats (Molloy 2010, 422-423). It is therefore possible that the set of values which were being constructed by some societies of the early Mycenaean Greek mainland included a strong component in which certain kinds of violent behaviour by males was considered acceptable.

---

[7] See Paul Greenhalgh (1973; 1980) and Drews (1993) for an alternative view.
[8] Paul Rehak (1995, 114 and n.206) notes the association of this iconography with female rather than male burials, raising the question of the significance of violence to concepts of gender.
[9] Papadopoulos (2006) dates the Cretan examples to LM I.

Given that the creation of a scheme of value also creates power asymmetries and inequalities (Damilati and Vavouranakis 2011, 39), such values in the early Mycenaean Greek mainland may have had a significant impact upon the development of societies. Values are created and maintained by practice, and the practice of particular kinds of violence may have formed an important aspect of social change during this period.

If individualised confrontations involving swords did take place in the Greek mainland at this time, their significance was in expressing the social values which were simultaneously being constructed through other practices, such as mortuary deposition. While it would be unwise to conclude from weapon-burials that the deceased took part in violent activities during life (Whitley 2002), osteological examination suggests that some of those interred in the Mycenae Grave Circles may have done so (Angel 1973, 393; Arnott 1999, 500; Voutsaki et al. 2007, 90-91). The use of swords in violence was only one aspect, however significant, of the deployment of material culture to create and express a new scheme of social values.

The social dynamic of the Shaft Graves and other elaborated mortuary practices of the early Mycenaean period is increasingly seen as one in which power relations were diffuse, rather than those of a 'royal', 'princely', or 'chiefly' society suggested by previous generations of scholars. Power should not be seen as entirely fluid, however. While the ability to obtain resources and use them strategically for the creation of schemes of value and power was crucial to the creation of new forms of society, the ability to participate in such competition was not open to all. That the mortuary practices of MH III/LH I demonstrate many kinds of continuity with those of the preceding MH period (Voutsaki 1999, 105-106) may suggest that there was continuity in terms of the structuring of access to power. Nor should it be assumed that political power directly correlates with the distribution of 'elite' material culture (Adams 2004, 193). Wolpert's (2004, 134) description of the Kwakiulti *potlatch* ceremonies, in which status competition was limited to those coming from a permanent pool of 658 formal names and which could not be altered by material strategies, is instructive in this regard.

During LH II, the final phase of the early Mycenaean period, further changes can be seen to occur in mortuary practices, not least the adoption of the tholos tomb as an 'elite' form of burial outside its Messenian origin and the use of the chamber tomb by many other communities. This period also sees a comparative lack of sword deposition in burials, although the robbing of almost all tholoi in antiquity (Deger-Jalkotzy 2006, 152) means that this may be a problem of evidence rather than a shift in social practice.

The 'warlike Mycenaeans' concept has often been used to construct narratives in which mainland groups were the agents of the destructions and social changes occurring on Crete in the LM IB-II periods (Caskey 1969, 442; Mellersh 1970, 122-123; Hood 1971; 1973; Doumas 1983, 146-147; Hood 1985; Driessen 1990, 117-125; Warren 1991; Driessen and Macdonald 1997). The apparent arrival of mainland funerary practices, and in particular the interment of weaponry in 'warrior graves', appeared to lend credence to these narratives. However, the excavation at the cemetery of Poros, a harbour-town close to the 'palace' at Knossos, yielded burials with swords/daggers, spears, and boar's tusk helmets dating to LM IA, prior to the destruction of the Neopalatial 'palaces' in LM IB (Dimopoulou 1999). Furthermore, strontium isotope analysis of skeletal material from LM II-IIIA burials in the vicinity of Knossos produced results which are not compatible with the sampled individuals having come from Myceane in the Argolid, but which are similar to those of MM III-LM I burials from the region (Nafplioti 2008). These findings suggest that the practice of weapon-interment was a feature of Neopalatial funerary ritual on Crete, and that changes in mortuary behaviour in LM II-IIIA may be best explained without recourse to invasion by mainlanders (see also Molloy (this volume)).

Palatial Period: Late Helladic IIIA-IIIB (c.1400-1200 BC)

The transition to the 'palatial' period proper is marked by the construction of megastructures (referred to as 'palaces') at Mycenae, Tiryns, and Midea in the Argolid, and at Ano Englianos (Pylos)[10] in Messenia during LH IIIA1. Further complexes were constructed at Thebes, Gla, Dimini, Athens, and Orchomenos (Spyropoulos 1974; Iakovidis 1983; Dakouri-Hild 2001; Adrimi-Sismani 2006; Wright 2006; Fitzsimmons 2011) during LH IIIA-B. That period also saw significant buildings constructed at Menelaion, Kanakia on Salamis, Agios Vasileios in Laconia, and Nichoria (Aschenbrenner et al. 1992, 433-439; Blackman 2000-2001, 14-15; 2001-2002, 14-15; Whitley 2004-2005; Catling 2009, 32-54; Cavanagh 2010-2011, 23). These structures are generally seen as being subordinate to the 'palaces', or as representing a lower level of social complexity, although Guy Middleton (2010, 5) notes the problematic nature of imposing such a simple division onto the archaeological record. Major settlements without significant buildings also exist, such as Teikhos Dymaion and Pellana (Papadopoulos 1979, 24; Spyropoulos 1998; Gazis 2010, 242-244).

The construction of the mainland 'palaces' has often been viewed as a signifier of a neoevolutionary shift to a 'state' society (Renfrew 1972 [2011], 369; Shelmerdine 2001, 349). This classification characterises Mycenaean 'palatial' societies as centralised hierarchies centred on the 'palaces'. This interpretation can be seen in Moses Finley's (1956; 1981) reconstruction of the 'palatial' economies as redistributive (i.e. dominated by a powerful central state which controlled the majority of economic

---

[10] Although substantial constructions at Ano Englianos (Pylos) predate LH IIIA (Nelson 2001, 194-200).

activity), and in Klaus Kilian's (1987, 204-205) view of the 'palace' as the central headquarters of political, legal, military, economic, and religious power. The social structure of Mycenaean society is generally envisaged as constituting a roughly pyramidal hierarchy, with the *wa-na-ka* or *wanax* ('lord', 'master') of the Linear B texts functioning as a monarchical figure at the apex, surmounting various groups of officials and administrators (Shelmerdine 2008, 127-135).

A full critique of this reconstruction of Mycenaean social structure is a task which must wait for a future paper, but a number of points may be made here. The reconstruction rests largely upon the Linear B texts discovered at Ano Englianos (Pylos) which constitute the largest proportion of surviving texts from the mainland. Therefore, the reconstruction may be viable for Messenia but not for other areas of the Greek mainland. The work of Paul Halstead (1992a; 1992b) has effectively dispelled Finley's vision of a centralised redistributive economy controlled by the 'palaces'. The identification of the *wanax* as the preeminent figure rests upon the fact that he is listed in the Pylian texts as having the largest landholding (Shelmerdine 2008, 128) and on his apparent centrality to the matters recorded in the texts (Palaima 2006, 64-68). This constitutes an explicit equation between land-wealth and power which is not necessarily apposite, while the centrality of the *wanax* in Linear B documents is questionable (Shelmerdine 2008, 128), and even if true may be hostage to power-strategies limiting the use of writing (e.g. Blanton 1998, 161). For a society organised upon individualising, 'executive' lines (Wright 2006, 161 and n.11), Mycenaean 'palatial' society exhibits little in the way of ruler iconography. Alternative reconstructions evoking a more corporate and heterarchical society are certainly possible, as Rehak (1995, 116) demonstrates:

> Even in the tablets, Mycenaean society seems to have been based on several interrelated figures of authority who represented different power bases: the *wanax* (a borrowed, non-Indo-European word and concept), the *lawagetas* (based transparently on the *laos*, or people, perhaps marshalled as a military host), the *damokoro* (from the *damos* or *demos*), and *potnia* ("she who has power" in Greek, probably a secular title, not a divine name). The picture that emerges from the tablets is one of a system of authority based on several sources that were gradually but perhaps unevenly integrated over time.

The classification of the Mycenaean 'palaces' as the cores of conventional states has, additionally, had an effect upon the study of warfare in the 'palatial' period of the Greek mainland, due to the continuing influence of the Weberian definition of the state as an entity which maintains a monopoly on legitimate violence (Weber 1947, 143). This model has led to a militaristic view of the 'palace' societies of LH IIIA-B (e.g. Bennet and Davis 1999; Deger-Jalkotzy 1999; Palaima 1999). The Weberian definition of the state – explicitly constructed with reference to the modern state – has been found increasingly at odds with the testimony of historical sources (Mann 1986, 11; Richardson 2012), rendering *a priori* assumptions about the nature of violence in 'state' societies problematic. An examination of the evidence for warfare and its relationship with society on the 'palatial' Greek mainland should therefore proceed with a more critical approach than has sometimes been seen in the past.

That there was an effort on the part of human agents on the Greek mainland to associate certain kinds of violence with the 'palatial' megastructures – or at least with parts of those megastructures – cannot be doubted. Figural frescoes depicting armed inter-human violence are present at Mycenae, Tiryns, Ano Englianos (Pylos), and at Orchomenos in LH III A-B, although the majority of these date to the latter part of the period (Immerwahr 1990, 123-128; Papadopoulos 2006). Scenes of inter-human violence on portable glyptic or ceramic media are rarer than in the early Mycenaean period, and a rise in the depiction of armed figures on pictorial ceramics towards the end of the 'palatial' period emphasises their comparative absence for much of LH IIIA-B.

In the mortuary sphere, the interment of weaponry with burials continued into LH IIIA1, but deposition is somewhat reduced, with the interment of multiple swords now being practiced less (Kilian-Dirlmeier 1993, abb. 28-29). Burials with armour appear confined to LH IIIA, with examples noted at Dendra (Åström 1977), Nichoria (Wilkie and Dickinson 1992, 276-278), and Mycenae (Catling 1977b, 102). Burials with military equipment do not appear to be centralised at 'palatial' sites during this period. This may be an artefact of post-depositional processes, in that all elite tholos tombs at 'palatial' sites have been looted, preventing us from assessing the role of militaristic items in their funerary assemblages (Deger-Jalkotzy 2006, 152). The presence of burials with military equipment need not relate to actual violent practices, although the decision to inter, or not inter, such material in graves forms a component of the creation and maintenance of social values.

The presence of documents written in the Linear B script provides us with a further source of evidence regarding violence in the 'palatial' period, although the concentration of these documents at 'palatial' sites, the fact that the vast majority of mainland examples are from the single site of Ano Englianos (Pylos), and their late date (LH IIIB2) render them problematic. The Linear B texts from Ano Englianos (Pylos) and Tiryns record the manufacture or storage of quantities of chariots, bronze body armour, swords, spears, and bows (Palaima 1999, 368; Bernabé and Luján 2008, 206-210 and 213-217). The Pylian archive provides further insight into military organisation, with the *o-ka* texts (part of the PY An-series) recording the names of officers and the number of men under their command, along with the locations of these units (Ventris and Chadwick 1973, 185).

These textual sources have been used to argue for the mainland 'palaces' possessing a strong, centralised military force (Lejeune 1972, 73-77; Palmer 1977, 252-253; Godart 1987; Deger-Jalkotzy 1999, 124-125; Palaima 1999, 367-369). This may be misleading, however. While it must be borne in mind that the Linear B texts are only a fragmentary sample of what must once have existed, it is also noteworthy that they appear to be biased towards recording particular classes of military equipment. Bronze body armour and chariots are recorded in numbers at Pylos, while swords and spears are much more infrequent, despite being a far more common item in the archaeological record.[11] The documented arsenals of the 'palaces' appear to have been more concerned with rarer and more exclusive armaments rather than the equipping of rank-and-file troops.[12] Archaeologically, the bronze corslet type of armour found in a chamber tomb at Dendra is thereafter limited to 'palatial' contexts, in the form of two examples from 'palatial' Thebes dating to LH IIIA/B and LH IIIB1 (Andrikou 2007). It is also unclear whether chariots in an Aegean context represent a vehicle of great military effectiveness (Littauer and Crouwel 1983; 1996).

As regards military organisation in the Linear B texts, Shelmerdine (2006, 79) has noted the relative lack of reference to military organisation in the extant corpus of texts, along with the lack of reference to military rations, bedding and other logistics which are documented for some types of industrial workers. Similarly, the *o-ka* tablets record the numbers and locations of units of men, with no indication of a command structure above that of the named commanders. The implication of this is that the 'palatial' bureaucracy – at least at Pylos, where more detailed textual evidence is available – did not exert a close level of organisation or control over military forces. A decentralised organisation of military forces, in which bodies of men were raised and equipped by local communities or landowners, and placed at the disposal of the 'palatial' bureaucracy as required, is consistent with other aspects of Pylian 'palatial' operations as illustrated by the Linear B texts (Halstead 1992a; 1992b; Shelmerdine 2006). Anthropologically, the presence of what may be termed a 'state' does not necessarily correlate to the presence of a strong centralised military force (Southall 1999, 33).

In the decentralised view of 'palatial' military organisation adopted above, a particular consequence would be the importance it placed on the maintenance of relationships between the 'palaces' and the communities and landowners upon which they relied for the greater part of their military forces. In such a context, the shifts in iconographic and mortuary practices away from portable items which could be associated with particular individuals and towards the more monumental (e.g. frescoes), may be read as a strategy to promote an understanding of the world in which particular forms of violence were associated with the 'palace' and those connected with it. The centralising nature of this ideology may have been at odds with – and, indeed, counterpoint to – the facts on the ground.

Material ties were doubtless important in constructing and maintaining networks of relationships in a variety of contexts. The ability of a 'palace' to grant holdings of land to an individual in return for service (Shelmerdine 2006, 75) may well have carried an obligation to future service, including the military service described by Chadwick (1987), which may have been met by landowners through the provision of a contingent of men rather than serving themselves. The equipping of elite combatants by the 'palatial' arsenals provides a direct link between the 'palatial' bureaucracies and the maintenance of relationships with local elites. The Knossian Linear B texts include chariot wheels named after individuals designated as *e-qe-ta* ('companions'), people who are elsewhere listed as the owners of slaves and textiles (Ventris and Chadwick 1973, 429), apparently demonstrating the equipping of high-status individuals with chariots and bronze body armour by the 'palace'. The presence of chariot wheels and body armour in the Pylian texts may suggest a similar dynamic in operation at that site.

The ideological basis of these relationships may be observed in 'palatial' art. Bennet and Davis (1999, 113) have argued that the Hall 64 fresco at Ano Englianos (Pylos), which shows a combat between animal skin-clad 'barbarians', and 'Mycenaeans' in boar's tusk helmets, is designed according to the ideology that to be Mycenaean is to be part of the 'palatial' system. Given the violent nature of the depiction, this may be extended to suggest that the 'Myceneaness' being promoted was also connected with the concept of performing military service to the 'palace'. The suggestion that the Hall 64 frescoes would serve as a backdrop to feasting activities taking place in Court 63 (Bennet and Davis 1999, 110) is also of interest, as feasts represent communal activities at which a variety of social relationships could be constructed and maintained.

Although it is common to treat 'the palace' as a single, unified entity, such an interpretation is not necessary. While social relationships structured by warfare doubtless did exist between actors located at the 'palace' and those located elsewhere, a more heterarchical approach to power, such as that outlined above, would result in the position of military power within the 'palace' itself being more complex than has previously been allowed for. The locations of violent iconography within the megastructures of the Greek mainland, for example in the main 'megaron' at Mycenae, but in the 'Southwestern Building' at Ano Englianos (Pylos),[13] may relate to the

---

[11] In 2009 excavations at Agios Vasileios in Laconia recovered not only a number of bronze swords, but also Linear B tablets, one of which recorded 500 bronze daggers (Cavanagh 2010-2011, 23).

[12] Linear B tablets from the Final Palatial period at Knossos on Crete may show the bureaucracy there equipping men of lower rank (Driessen and Macdonald 1984, 55-56). Even if this is so, Knossos does not necessarily provide a model which should be applied to all other 'palaces'.

[13] Lisa Bendall (2003) argues that the Northeastern building at Ano Englianos (Pylos) was a storage and administration facility with a strong connection to military matters.

differing relationships of power which existed in different megastructures, and perhaps also to different social contexts for violence in different regions of the Greek mainland.

Significant changes occur to the military equipment used in the Greek mainland during the LH IIIB2 period. The most obvious changes are the arrival of the Naue II sword which had originally developed in central Europe, the 'flame-shaped' spearheads which Anthony Snodgrass (1964, 119), saw as having strong parallels in central Europe,[14] and the reappearance – after an absence of c.150 years – of the bronze greave in the Greek mainland.[15] While Cyprus cannot in any sense be considered part of the Aegean, the reappearance of metal greaves in the Aegean is attested by the find of a fragment of a greave of Aegean manufacture from Tomb 18 of the Swedish excavations at Enkomi, Cyprus, dating to Late Cypriot IIC (equivalent to LH IIIB) (Catling 1955, 35; Fortenberry 1991, 626). The wire lacing of the Enkomi greave suggests a link to bronze greaves from Italy and the Balkans from perhaps as early as the 13th century BC (Clausing 2002, 150-164). A small round shield may have come into use in the Aegean during this period (Fortenberry 1990, 23-26), although the evidence for it is purely iconographic.

The presence of these items in the Aegean during the later part of the 'palatial' period and their connections to Italy and central/south-eastern Europe, in combination with the destruction of the 'palaces' at the end of that period, has led numerous scholars to conclude that there was an 'invasion' of or 'migration' to the Aegean by people from elsewhere, which resulted in the destruction of the 'palaces' and the introduction of new cultural elements (Desborough 1964; Sandars 1964; Vermeule 1964; Drews 1993; Bouzek 1994). A migration-based explanation for the end of the 'palatial' megastructures is archaeologically problematic, however. Several 'new' items of material culture appeared in the 'palatial' heartlands at a point considerably before the destructions.[16] There are more general problems arising from the assumption that items of material culture carry some form of 'ethnic identity', rather than being easily transferrable across human social boundaries. The arrival of new military items in the Aegean towards the close of the 'palatial' period should therefore be considered to represent a change in Mycenaean ideologies and their attendant practices, rather than a change in population.[17]

The destructions of the 'palaces', and other sites at the LH IIIB/C transition (c.1200 BC), are often classed as anthropogenic in nature, although other explanations have been put forward. Warfare may form one component of those events which characterised the end of the 'palatial' period, although its role must not be over-estimated, as the concept of a 'military revolution' which swept away the 'palatial' militaries (Drews 1993) is unsustainable given the archaeological evidence (see Littauer and Crouwel 1996; Dickinson 1999). Even a brief survey of the anthropological literature on warfare demonstrates that the motivations for engaging in violence are many and varied, and considerations of that spectrum of reasons must form part of our explanation. Totalising explanations for the end of 'palatial' society should be avoided, as many previous explanations have foundered on their assumption that a single phenomenon must account for those changes which are archaeologically visible. The apparent social and political disparities in different regions during the LH IIIC period suggest that very different sequences of events may have occurred in different areas of the mainland in the closing years of the 'palatial' period. While archaeological sources are not suitable for constructing the text-based histories that have been attempted for other regions (such as that accomplished by Trevor Bryce (1998) for the Hittite empire), an awareness of change as an historical process must inform our explanations.

Postpalatial Period: Late Helladic IIIC (c.1200-1100 BC)

The destruction of the 'palatial' megastructures at the end of LH IIIB has been seen by many scholars as representing devolution in the level of social complexity on both the Greek mainland and across the wider Aegean world (e.g. Starr 1977, 47; Deger-Jalkotzy 2008, 392; Middleton 2010, 31). However, even if architecture was to be the sole measure of 'social complexity' then, as has been noted previously, a number of significant 'non-palatial' structures existed during the 'palatial' period. Significant structures continue to exist in the postpalatial period, with examples at Mycenae, Tiryns and Midea in the Argolid, and at Pyrgos Livanaton and Mitrou in east Lokris (Walberg 1995; Maran 2001; French 2002, 136-138; Dakoronia 2003, 38; Maran 2006; van de Moortel 2009, 361-362), while a major settlement existed at Xeropolis-Lefkandi on Euboia (Sherratt 2006). In some respects, therefore, there would seem to have been substantial continuity between 'palatial' and postpalatial periods. Given what has already been said regarding different forms of social complexity and the fact that not all may find material expression in the archaeological record, then while the LH IIIB-C transition was clearly a time of significant changes in the Greek mainland, these changes are not, necessarily, best understood as shifts in vertical position on an evolutionary scale. It has been common to view postpalatial and Early Iron Age societies in the Aegean either as aristocratic 'chiefdoms' (Snodgrass 1971 [2000], 386-388; Ferguson 1991; Dickinson 2006, 110-112) or as 'big man' societies analogous to those described in the anthropology of Melanesia (Sahlins 1963; Donlan 1985; Murray 1993, 46-48; Hall 2007, 120-127; Middleton 2010, 94-97). While such social formations may indeed have existed in

---

[14] Although Molloy (personal communication) sees more frequent continuity with local spearhead forms.
[15] Iconographic evidence of the LH IIIA-B period does appear to show the use of leg protectors (Catling 1977a, 150-152; Fortenberry 1991), but the material these items were made from is difficult to ascertain.
[16] The first Aegean evidence of the Naue II sword coming in the form of an ivory hilt plate from Mycenae (Krzyszkowska 1997, 147).
[17] Middleton (2010, 73) provides a useful insight into how population movement may be understood in past contexts.

particular times and at particular places during the period in question, they were not necessarily the *only* kinds of social formation which existed (Whitley 1991, 191-194; Foxhall 1995; Haggis 1999).

The material evidence for the practice of warfare in the postpalatial Greek mainland exhibits a great deal of continuity with that of the last phase of the 'palatial' period. The boar's tusk helmet survives until the end of the period,[18] but is now supplemented with a number of other helmet-types (Borchhardt, J. 1977, 66-68; Dickinson 2006, 74), the Naue II sword is used alongside the Aegean Type F and G swords until the end of the period (Kilian-Dirlmeier 1993, 106-115), a number of different types of shield appear in iconographic sources (Borchhardt, H. 1977), and body armour is attested in a number of forms (Catling 1977b; Maran 2004, 18-24). These changes may be significant in that the items which go out of use appear to be those particularly associated with 'palatial' culture.[19] The use of alternative items, whether developed locally or derived from equipment used in other geographic regions, should be seen as symptomatic of those social processes surrounding the destruction and non-replacement of the 'palatial' megastructures.

The broad continuity of military equipment between the LH IIIB and LH IIIC periods suggests that the removal of the 'palaces' had little effect upon how fighters were armed. Whether there were significant changes to the way in which armed violence was practiced is more debatable. Based upon the study of military equipment, Molloy (2010, 423-424) sees the tactics of this period as shifting to a more cooperative and organised style of fighting, with the thrusting spear as the primary weapon. The smaller size of spearheads in LH IIIB-C and iconographic evidence of combatants carrying two spears may indicate the use of the spear as a multipurpose weapon for throwing or thrusting (Höckmann 1980, 111-112). This comparatively unregimented mode of warfare may have persisted down to at least the Archaic period, given literary and iconographic evidence (van Wees 2004, 166-170).

The absence of the 'palaces' in this period might be expected to have produced significant changes to the organisation and equipping of military forces, but this does not appear to have been the case. The chariot, which would have been both difficult to manufacture and expensive to maintain, is manifest in the pictorial kraters of LH IIIC Middle (Crouwel 1999). Swords are more plentiful in the postpalatial archaeological record than in that of the 'palatial' period, and while this is doubtless partly due to different social restrictions being placed upon their deposition (Molloy 2010, 423), it suggests that there was no great shortage of them in the absence of 'palatial' modes of manufacture. The continuity of organisation between 'palatial' and postpalatial periods may be explained by the decentralised organisation of warfare argued for above, in which 'palatial' military forces were primarily equipped and organised by local authorities. These methods would be resilient to the disappearance of a central authority which had never played a major role in them. In the postpalatial period, military forces would still be armed and organised by local authorities. The major change resulting from this would be a potential reduction in the size of the largest military force which could be mustered by any one authority, with smaller forces being the norm, but the units of men recorded in the Pylian *o-ka* texts may represent an appropriate measure of the military forces of the postpalatial period.

The divergent social formations which characterised different regions of the Greek mainland in LH IIIC gave rise to different uses of violence-related material culture in the creation and maintenance of social values. Due to the nature of the postpalatial archaeological record: not all regions of the Greek mainland offer a suitable evidential base for study. Messenia, for example, appears to have encountered a significant fall in the population level in LH IIIC (Davis et al. 1997, 424), resulting in a relatively insubstantial archaeological record. This is unfortunate given the wealth of evidence that the region yielded for the 'palatial' period, as it makes a discussion of the development of the region almost impossible. Discussion will therefore be limited to regions which provide suitable assemblages, with an emphasis on understanding the material in the context of local social formations.

The Argolid provides a particularly interesting case study in the postpalatial period. Mycenae, Tiryns and Midea – all three the sites of 'palatial' period megastructures – saw reoccupation during LH IIIC. The area of the former megaron at Tiryns saw the construction of 'Building T' in the postpalatial period, while a similar structure was erected at Midea, and a building below the Archaic temple at Mycenae, also located on the site of the former megaron, may date to LH IIIC (Maran 2006, 124-125). In the case of Tiryns the excavator views 'Building T' as a ceremonial structure, with houses significant in terms of size and ground plan existing in the Lower Town below the citadel (Maran 2006, 125-126). The nature of the buildings at Midea and Mycenae is more difficult to determine, but at Mycenae there are also three LH IIIC buildings (the last featuring fresco compositions) on the site of the 'palatial' period 'House Alpha' (Maran 2006, 143). The find next to 'Megaron W' of the 'Tiryns Treasure', an LH IIIC deposition of valuable items (some dating as far back as the Early Mycenaean period), enhances the impression that in the Argolid, postpalatial political formations were constructed with explicit reference to the 'palatial' past.

---

[18] Cynthia Shelmerdine (1996) argues that the use of boar's tusk helmets as physical items (rather than as an iconographic element) diminishes in LH IIIA-B. The latest find of a boar's tusk helmet appears to be that from an LH IIIC Middle or Late weapon-burial at Kallithea-Spenzes in Achaea (Deger-Jalkotzy 2006, 160-161).

[19] Both the boar's tusk helmet and the figure-of-eight shield remained in use – at least iconographically – in LH IIIA-B (Immerwahr 1990, 138-140; Papadopoulos 2006).

Notable in the archaeological record of the postpalatial Argolid is the presence of violence-related iconography in the form of the 'Warrior Vase' and 'Warrior Stele' from Mycenae and the pictorial kraters and *pinax* or *larnax* fragment from Tiyns which feature armed charioteers (Immerwahr 1990, 149-153). These depictions perhaps follow 'palatial' norms in showing groups of undifferentiated fighters rather than individualised confrontations. The use-contexts of the ceramics are unclear, although a funerary purpose is possible. The use of the 'Warrior Stele' at the entrance to a chamber tomb (Vermeule and Karageorghis 1982, 132) may have served a similar purpose to sword-interment,[20] but with the context of the chamber tomb directing its focus as onto a collective group rather than any particular individual.

While the archaeological record of the postpalatial Argolid does not appear to support the presence of 'chiefdoms', nor does it possess the ephemerality which might be expected of a society based around 'big men' (Dickinson 2006, 110-111). At both Mycenae and Tiryns there is a more diffuse distribution of elite material culture, and power may have exhibited a similar distribution.

A different set of practices is found in Achaea. No significant postpalatial buildings have been discovered in the region, although the settlement at Teikhos Dymaion was occupied between destructions in LH IIIB-C and at the end of LH IIIC (Papadopoulos 1979, 24; Gazis 2010, 242). Achaea has not produced any examples of violence-related iconography which are comparable to those from the Argolid in the same period.

Achaea has produced at least 11 'warrior burials' dating to LH IIIC, far surpassing the number from any other region of the Aegean in this period. Even allowing for differential levels of preservation and archaeological investigation between regions, it is clear that interment with weapons was a practice of particular significance in this region during the postpalatial period. In Achaea the values associated with swords and other military equipment were expressed in a funerary format. While this may be consistent with the 'big man' concept of short-lived groupings forming around particular individuals, such a direct reading of the archaeological record is not unproblematic.

A further significant factor in the development of weapon-burials in Achaea was the location of the region close to Adriatic routes of contact and exchange. A number of items found in the weapon-burials of Achaea and other regions of western Greece have strong connections to Italy, particularly swords and metal greaves (Eder and Jung 2005). Weapon-burial was also a practice in areas of the Italian peninsula during the 12th century BC, suggesting possible similarities of practice on both sides of the Adriatic. It may be more productive to consider Achaean societies as belonging to an 'Adriatic' world rather than a 'Aegean' one, for the purposes of analysis.[21]

A third region suitable for study here is that of the Euboean Gulf. During LH IIIC the region saw significant building-complexes constructed at Pyrgos Livanaton and Mitrou in east Lokris (Dakoronia 2003, 38; van de Moortel 2009, 361-362), and a major settlement developing at Xeropolis-Lefkandi on Euboia (Sherratt 2006). Unlike Achaea, the practice of weapon-interments is not attested in the Euboian Gulf during LH IIIC. At Pyrgos Livanaton, a new form of violent iconography is found, with painted ceramics featuring scenes of armed men on ships (Dakoronia 1987; 1996; 1999). This iconography is substantially different to that of the Argolid in the same period, but also presents a change to the Aegean artistic repertoire in that it appears to show naval battles in the open water for the first time (Papadopoulos 2006, 232-233). Given the coastal location of Pyrgos Livanaton, this imagery reflects the importance of maritime activities in the lives of inhabitants. 'Warrior iconography' similar to that of the Argolid is found on ceramics from the site of Xeropolis-Lefkandi (Crouwel 2006, 238-240). Centralised, hierarchical and 'individualising' relations of power may be appropriate for some, but not necessarily all, of these sites.

This LH IIIC period sees a resurgence in the practice of weapon-burial in the Greek mainland compared to LH IIIA-B, and it is tempting to see this as the development of a form of 'inversion' (Miller 1989, 66-68) in which a set of values antithetical to those which are dominant is created from the same set of ideas. This may relate to the end of the 'palatial' megastructures and the formation of new social structures. Such developments, and any attendant presence of physical violence in society, do not, however, appear to be associated with any particular socio-political formation, and it is furthermore unclear whether the values being expressed were in fact the same in each region.

An Aegean 'ethnicity' for at least some of the 'Sea Peoples' recorded by the Egyptians, and said by them to be responsible for raids and destructions in the Eastern Mediterranean, has been mooted by numerous scholars (Sandars 1978; Dothan 1982; Dothan and Dothan 1992; Stager 1998; Yasur-Landau 2010; Kaniewski et al. 2011). Some connection is possible given the similarity between the distinctive headgear or hairstyle sported by the 'Sea Peoples' in Egyptian and Cypriot depictions, and by some figures in Aegean and Anatolian iconography. However, attempting to prescribe an 'ethnic identity' based on archaeological evidence is problematic, as concepts of ethnicity are often more complex and

---

[20] This is similar to the suggestion by Oliver Dickinson (2006, 194) that the comparative absence of weapon-burials in 8th century BC Athens may be offset by the presence of 'warrior' iconography on ceramics used in funerary contexts.

[21] The concept of network theorists such as Irad Malkin (2011, 9) that connectivity is more important than geographical distance may be apposite here.

ambiguous than allowed for by simplistic modelling (Hall 1995). The distinctive headgear of the 'Sea Peoples' may have been a fashion which transcended cultural boundaries in the Eastern Mediterranean, or may have served as the marker of a piratical subculture rather than of an 'ethnicity'. Commonalities of material culture do not necessarily suggest large scale population movements, and consequently it would be erroneous to conclude that the 'Sea Peoples' represent a violent migration of Aegean origin, or of Aegean-based military activities in the wider region.

Conclusion

The dual approach taken here in the examination of warfare and social complexity on the Late Bronze Age Greek mainland, has produced insights into both. Critiquing neoevolutionary typologies which would see the societies of the region as becoming more centralised and hierarchical allows us to consider that power – including violent power – may have been distributed in a diffuse and uneven way. This in turn suggests that violence should not be viewed simply as a method of integration used by centralising polities, but rather as something which could potentially be deployed for various purposes by a number of agents.

The decoupling of forms of warfare from particular evolutionary 'levels', and the multiplicity of forms of social organisation that this paper has suggested for the Greek mainland, further suggests that the nature of violence and the social uses to which violence and violent concepts were put will be best understood by the close examination of the archaeological record in a given area. The examination of the use of violent concepts in the construction of social values, as has been done in a preliminary fashion in this paper, is one method by which such future work may be conducted.

Acknowledgements

My sincere thanks are due to Dr David Smith (University of Liverpool), Dr Barry Molloy (University of Sheffield), Dr Kate Harrell (Polk State College), and Professor Christopher Mee (University of Liverpool) for the assistance and constructive criticism they provided for this paper. Any remaining errors or misinterpretations must surely be my own.

Bibliography

Adams, E. 2004. Power Relations in Minoan Palatial Towns: An analysis of Neopalatial Knossos and Malia. *Journal of Mediterranean Archaeology* 17(2), 191-222.

Adrimi-Sismani, V. 2006. The Palace of Iolkos and Its End. In S. Deger-Jalkotzy and I. S. Lemos (eds.), *Ancient Greece: From the Mycenaean palaces to the age of Homer*. Edinburgh Leventis Studies 3, 465-483. Edinburgh, Edinburgh University Press.

Allen, M. W. and Arkush, E. N. 2006. Introduction. In E. N. Arkush and M. W. Allen (eds.), *The Archaeology of Warfare: Prehistories of raiding and conquest*. 1-19. Gainesville FL, University Press of Florida.

Andrikou, E. 2007. New Evidence on Mycenaean Bronze Corslets from Thebes in Boeotia and the Bronze Age Sequence of Corslets in Greece and Europe. In I. Galanakis, H. Tomas, Y. Galanakis and R. Laffineur (eds.), *Between the Aegean and Baltic Seas: Prehistory across borders*. Aegaeum 27, 401-411. Liège, Université de Liège.

Angel, J. L. 1973. Human Skeletons from Grave Circles at Mycenae. In G. E. Mylonas, *Ο Ταφικός Κύκλος Β των Μυκηνών*. Vol. I. 379-397. Athens, Archaeological Society of Athens.

Arnott, R. 1999. War Wounds and Their Treatment in the Aegean Bronze Age. In R. Laffineur (ed.), *POLEMOS: Le contexte guerrier en Égée à l'âge du Bronze*. Vol. II. Aegaeum 19, 499-507. Liège, Université de Liège.

Aschenbrenner, S. E., Coulson, W. D. E., Donovan, W. P., Hope Simpson, R., McDonald, W. A. and Wilkie, N. C. 1992. Late Helladic Settlement: Stratigraphy and architecture. In W. A. McDonald and N. C. Wilkie (eds.), *The Excavations at Nichoria in Southwest Greece Vol. II: The Bronze Age occupation*. 359-454. Minneapolis MN, University of Minnesota Press.

Åström, P. 1977. *The Cuirass Tomb and Other Finds at Dendra Part I: The chamber tombs*. Studies in Mediterranean Archaeology 4. Gothenburg, Paul Åströms Förlag.

Avila, R. A. J. 1983. *Bronzene Lanzen- und Pfeilspitzen der griechischen Spätbronzezeit*. Prähistorische Bronzefunde V,1. Munich, C. H. Beck'sche.

Bendall, L. M. 2003. A Reconsideration of the Northeastern Building at Pylos: Evidence for a redistributive centre. *American Journal of Archaeology* 107(2), 181-231.

Bennet, J. and Davis, J. L. 1999. Making Mycenaeans: Warfare, territorial expansion and representations of the other in the Pylian kingdom. In R. Laffineur (ed.), *POLEMOS: Le contexte guerrier en Égée à l'âge du Bronze*. Vol. I. Aegaeum 19, 105-121. Liège, Université de Liège.

Bernabé, A. and Luján, E. R. 2008. Mycenaean Technology. In Y. Duhoux and A. Morpurgo Davies (eds.), *A Companion to Linear B: Mycenaean Greek texts and their world*. Vol. I. Bibliothèque des Cahiers de l'Institut de Linguistique de Louvain 120, 201-233. Louvain-la-Neuve, Peeters.

Blackman, D. 2000-2001. Archaeology in Greece 2000-2001. *Archaeological Reports* 47, 1-144.

Blackman, D. 2001-2002. Archaeology in Greece 2001-2002. *Archaeological Reports* 48, 1-115.

Blanton, R. E. 1998. Beyond Centralization: Steps toward a theory of egalitarian behaviour in archaic states. In G. M. Feinman and J. Marcus (eds.), *Archaic States*. 135-172. Santa Fe NM, School of American Research Press.

Blanton, R. E., Feinman, G. M., Kowalewski, S. A. and Peregrine, P. N. 1996. A Dual-Processual Theory for the Evolution of Mesoamerican Civilization. *Current Anthropology* 37(1), 1-14.

Borchhardt, H. 1977. Frühe Griechische Shildformen. In H.-G. Buchholz and J. Wiesner (eds.), *Archaeologia Homerica. Kriegswesen: Schutzwaffen und Wehrbauten*. Archaeologia Homerica Band 1, Kapitel E, Teil 1, 1-56. Göttingen, Vandenhoeck and Ruprecht.

Borchhardt, J. 1977. Helme. In H.-G. Buchholz and J. Wiesner (eds.), *Archaeologia Homerica. Kriegswesen: Schutzwaffen und Wehrbauten*. Archaeologia Homerica Band 1, Kapitel E, Teil 1, 57-74. Göttingen, Vandenhoeck and Ruprecht.

Bouzek, J. 1994. Bronze Age Greece and the Balkans: A review of the present evidence. *Annual of the British School at Athens* 89, 217-234.

Bryce, T. R. 1998. *The Kingdom of the Hittites*. Oxford, Clarendon Press.

Buchholz, H.-G. 1962. Die Pfeilglätter aus dem VI. Schachtgrab vom Mykene und die helladischen Pfeilspitzen. *Jahrbuch des Deutschen Archäologischen Instituts* 77, 1-58.

Buchholz, H.-G. (ed.). 1980. *Archaeologia Homerica. Kriegswesen: Angriffswaffen*. Archaeologia Homerica Band 1, Kapitel E, Teil 2. Göttingen, Vandenhoeck and Ruprecht.

Buchholz, H.-G. and Wiesner, J. (eds.). 1977. *Archaeologia Homerica. Kriegswesen: Schutzwaffen und Wehrbauten*. Archaeologia Homerica Band 1, Kapitel E, Teil 1. Göttingen, Vandenhoeck and Ruprecht.

Carman, J. and Carman, P. 2005. War in Prehistoric Society: Modern views of ancient violence. In M. Parker Pearson and I. J. N. Thorpe (eds.), *Warfare, Violence and Slavery in Prehistory*. British Archaeological Reports International Series 1374, 217-225. Oxford, BAR Publishing.

Carneiro, R. L. 1970. A Theory on the Origin of the State. *Science* 169, 733-738.

Caskey, J. L. 1969. Crises in the Minoan-Mycenaean World. *Proceedings of the American Philosophical Society* 113(6), 433-449.

Catling, H. W. 1955. A Bronze Greave from a 13th Century BC Tomb at Enkomi. *Opuscula Atheniensia* 2, 21-36.

Catling, H. W. 1977a. Beinschienen. In H.-G. Buchholz and J. Wiesner (eds.), *Archaeologia Homerica. Kriegswesen: Schutzwaffen und Wehrbauten*. Archaeologia Homerica Band 1, Kapitel E, Teil 1, 143-161. Göttingen, Vandenhoeck and Ruprecht.

Catling, H. W. 1977b. Panzer. In H.-G. Buchholz and J. Wiesner (eds.), *Archaeologia Homerica. Kriegswesen: Schutzwaffen und Wehrbauten*. Archaeologia Homerica Band 1, Kapitel E, Teil 1, 74-118. Göttingen, Vandenhoeck and Ruprecht.

Catling, H. W. 2009. *Sparta: Menelaion I: The Bronze Age*. Vol. I. British School at Athens Supplementary Volume 45.

Cavanagh, W. G. 2010-2011. The Greek Mainland in the Prehistoric Period. *Archaeological Reports* 57, 19-26.

Chadwick, J. 1987. The Muster of the Pylian Fleet. In P. H. Ilievski and L. Crepajac (eds.), *Tractata Mycenaea: Proceedings of the Eighth International Colloquium on Mycenaean Studies*. 75-84. Skopje, Macedonian Academy of Sciences and Arts.

Cherry, J. F. 1986. Polities and Palaces: Some problems in Minoan state formation. In C. Renfrew and J. F. Cherry (eds.), *Peer Polity Interaction and Socio-Political Change*. 19-45. Cambridge, Cambridge University Press.

Claessen, H. and Skalník, P. 1978. Limits: Beginning and end of the early state. In H. Claessen and P. Skalník (eds.), *The Early State*. 619-635. The Hague, Mouton.

Clausing, C. 2002. Geschnürte Beinschienen der späten Bronze- und älteren Eisenzeit. *Jahrbuch des Römisch-germanischen Zentralmuseums, Mainz* 49, 149-187.

Cohen, R. 1984. Warfare and State Formation: Wars make states and states make wars. In R. B. Ferguson (ed.), *Warfare, Culture and Environment*. 329-358. New York NY, Academic Press.

Crouwel, J. H. 1981. *Chariots and Other Means of Land Transport in Bronze Age Greece*. Allard Pierson Series 3. Amsterdam, Allard Pierson Museum.

Crouwel, J. H. 1999. Fighting on Land and Sea in Late Mycenaean Times. In R. Laffineur (ed.), *POLEMOS: Le contexte guerrier en Égée à l'âge du Bronze*. Vol. II. Aegaeum 19, 455-463. Liège, Université de Liège.

Crouwel, J. H. 2006. Late Mycenaean Pictorial Pottery. In D. Evely (ed.), *Lefkandi IV: The Late Helladic IIIC Settlement at Xeropolis*. British School at Athens Supplementary Volume 39, 233-255. London, British School at Athens.

Dakoronia, F. 1987. War-Ships on Sherds of LH IIIC Kraters from Kynos. In H. E. Tzalas (ed.), *Tropis II: 2nd international symposium on ship construction in antiquity*. 117-122. Delphi, Hellenic Institute for the Preservation of Nautical Tradition.

Dakoronia, F. 1996. Kynos.....Fleet. In H. E. Tzalas (ed.), *Tropis IV: 4th international symposium on ship construction in antiquity*. 159-173. Athens, Hellenic Institute for the Preservation of Nautical Tradition.

Dakoronia, F. 1999. Representations of Sea-Battles on Mycenaean Sherds from Kynos. In H. E. Tzalas (ed.), *Tropis V: 5th international symposium on ship construction in antiquity*. 119-128. Athens, Hellenic Institute for the Preservation of Nautical Tradition.

Dakoronia, F. 2003. The Transition from Late Helladic III C to the Early Iron Age at Kynos. In S. Deger-Jalkotzy and M. Zavadil (eds.), *LH III C Chronology and Synchronisms*. Veröffentlichungen der Mykenischen Kommission 20, 37-51. Vienna, Verlag der Österreichischen Akademie der Wissenschaften.

Dakouri-Hild, A. 2001. The House of Kadmos in Mycenaean Thebes Reconsidered: Architecture, chronology, and context. *Annual of the British School at Athens* 96, 81-122.

Damilati, K. and Vavouranakis, G. 2011. 'Society Against the State?' Contextualixing inequality and power in Bronze Age Crete. In N. Terrenato and D. C. Haggis (eds.), *State Formation in Italy and Greece: Questioning the neoevolutionist paradigm*. 32-60. Oxford, Oxbow Books.

Davis, J. L., Alcock, S. E., Bennet, J., Lolos, Y. G. and Shelmerdine, C. W. 1997. The Pylos Regional Archaeological Project, Part 1: Overview and the archaeological survey. *Hesperia* 66(3), 391-494.

Deger-Jalkotzy, S. 1999. Military Prowess and Social Status in Mycenaean Greece. In R. Laffineur (ed.), *POLEMOS: Le contexte guerrier en Égée à l'âge du Bronze*. Vol. I. Aegaeum 19, 121-131. Liège, Université de Liège.

Deger-Jalkotzy, S. 2006. Late Mycenaean Warrior Tombs. In S. Deger-Jalkotzy and I. S. Lemos (eds.), *Ancient Greece: From the Mycenaean palaces to the age of Homer*. Edinburgh Leventis Studies 3, 151-181. Edinburgh, Edinburgh University Press.

Deger-Jalkotzy, S. 2008. Decline, Destruction, Aftermath. In C. W. Shelmerdine (ed.), *The Cambridge Companion to the Aegean Bronze Age*. 387-415. Cambridge, Cambridge University Press.

Dentan, R. K. 1988. On Reconsidering Violence in Simple Human Societies. *Current Anthropology* 29(4), 624-636.

Desborough, V. R. d'A. 1964. *The Last Mycenaeans and Their Successors: An archaeological survey c.1200-c.1000 BC*. Oxford, Clarendon.

Dickinson, O. T. P. K. 1977. *The Origins of Mycenaean Civilisation*. Studies in Mediterranean Archaeology 49. Gothenburg, Paul Åströms Förlag.

Dickinson, O. T. P. K. 1989. 'The Origins of Mycenaean Civilisation' Revisited. In R. Laffineur (ed.), *Transition: Le monde égéen du Bronze moyen au Bronze récent*. Aegaeum 3, 131-136. Liège, Université de Liège.

Dickinson, O. T. P. K. 1994. *The Aegean Bronze Age*. Cambridge, Cambridge University Press.

Dickinson, O. T. P. K. 1999. Robert Drews' Theories About the Nature of Warfare in the Late Bronze Age Aegean. In R. Laffineur (ed.), *POLEMOS: Le contexte guerrier en Égée à l'âge du Bronze*. Vol. I. Aegaeum 19, 21-29. Liège, Université de Liège.

Dickinson, O. T. P. K. 2006. *The Aegean from Bronze Age to Iron Age: Continuity and change between the twelfth and eighth centuries BC*. London, Routledge.

Dimopoulou, N. 1999. The Neopalatial Cemetery of the Knossian Harbour-Town at Poros: Mortuary behaviour and social ranking. In *Eliten in der Bronzezeit: Ergebnisse zweier Kolloquien in Mainz und Athen*. Römisch-Germanisches Zentralmuseum Mainz Monographien 43,1, 27-36. Mainz, Verlag des Römisch-Germanisches Zentralmuseums.

Donlan, W. 1985. The Social Groups of Dark Age Greece. *Classical Philology* 80(4), 293-308.

Dothan, T. 1982. *The Philistines and Their Material Culture*. New Haven CT, Yale University Press.

Dothan, T. and Dothan, M. 1992. *People of the Sea: The search for the Philistines*. New York NY, Macmillan.

Doumas, C. G. 1983. *Thera: Pompeii of the ancient Aegean. Excavations at Akrotiri 1967-79*. London, Thames and Hudson.

Drews, R. 1993. *The End of the Bronze Age: changes in warfare and the catastrophe ca. 1200 BC*. Princeton NJ, Princeton University Press.

Driessen, J. 1990. *An Early Destruction in the Mycenaean Palace at Knossos: A new interpretation of the excavation field-notes of the south-east area of the West Wing*. Acta Archaeologica Lovaniensia Monographiae 2. Leuven, Leuven University Press.

Driessen, J. and Macdonald, C. F. 1984. Some Military Aspects of the Aegean in the Late Fifteenth and Early Fourteenth Centuries BC. *Annual of the British School at Athens* 79, 49-74.

Driessen, J. and Macdonald, C. F. 1997. *The Troubled Island: Minoan Crete before and after the Santorini eruption.* Aegaeum 17. Liège, Université de Liège.

Earle, T. K. 1997. *How Chiefs Come to Power: The political economy in prehistory.* Stanford CA, Stanford University Press.

Eder, B. and Jung, R. 2005. On the Character of Social Relations Between Greece and Italy in the 12th/11th C. BC. In R. Laffineur and E. Greco (eds.), *EMPORIA: Aegeans in the central and eastern Mediterranean.* Aegaeum 25, 485-497. Liège, Université de Liège.

Evans, A. J. 1928. *The Palace of Minos.* Vol. II,2. London, MacMillan.

Evans, A. J. 1935. *The Palace of Minos.* Vol. IV,2. London, MacMillan.

Ferguson, R. B. 1997. Violence and War in Prehistory. In D. L. Martin and D. W. Frayer (eds.), *Troubled Times: Violence and warfare in the past.* War and Society 3, 321-357. New York NY, Routledge.

Ferguson, Y. H. 1991. Chiefdoms to City-States: The Greek experience. In T. K. Earle (ed.), *Chiefdoms: Power, economy, and ideology.* 169-192. Cambridge, Cambridge University Press.

Finley, M. I. 1956. *The World of Odysseus.* London, Chatto and Windus.

Finley, M. I. 1981. *Economy and Society in Ancient Greece.* London, Chatto and Windus.

Fitzsimmons, R. D. 2011. Monumental Architecture and the Construction of the Mycenaean State. In N. Terrenato and D. C. Haggis (eds.), *State Formation in Italy and Greece: Questioning the neoevolutionist paradigm.* 75-118. Oxford, Oxbow Books.

Flannery, K. V. 1972. The Cultural Evolution of Civilizations. *Annual Review of Ecology and Systematics* 3, 399-426.

Fortenberry, D. 1990. *Elements of Mycenaean Warfare.* Unpublished PhD thesis, University of Cincinnati.

Fortenberry, D. 1991. Single Greaves in the Late Helladic Period. *American Journal of Archaeology* 95(4), 623-627.

Foxhall, L. 1995. Bronze to Iron: Agricultural systems and political structures in Late Bronze Age and Early Iron Age Greece. *Annual of the British School at Athens* 90, 239-250.

French, E. B. 2002. *Mycenae Agamemnon's Capital: The site in its setting.* Stroud, Tempus.

Fried, M. H. 1961. Warfare, Military Organization, and the Evolution of Society. *Anthropologica* n.s. 3(2), 134-147.

Fried, M. H. 1967. *The Evolution of Political Society: An essay in political anthropology.* New York NY, Random House.

Gat, A. 2006. *War in Human Civilization.* Oxford, Oxford University Press.

Gauß, W., Lindblom, M. and Smetana, R. 2011. The Middle Helladic Large Building Complex at Kolonna. A preliminary view. In W. Gauß, M. Lindblom, R. Angus K. Smith and J. C. Wright (eds.), *Our Cups Are Full: Pottery and society in the Aegean Bronze Age. Papers presented to Jeremy B. Rutter on the occasion of his 65th birthday*, 76-87. Oxford, BAR Publishing.

Gazis, M. 2010. Η προϊστορική ακρόπολη του Τείχους Δυμαίων. Σε αναζήτηση ταυτότητας. In N. Merousis, E. Stefani and M. Nikolaidou (eds.), *ΙΡΙΣ. Μελέτες στη μνήμη της καθηγήτριας Αγγελικής Πιλάλη-Παπαστερίου.* 237-255. Thessaloniki, Editions Kornilia-Sfakianaki.

Georgousopoulou, T. 2004. Simplicity vs Complexity: Social relationships and the MHI community of Asine. In J. C. Barrett and P. Halstead (eds.), *The Emergence of Civilisation Revisited.* Sheffield Studies in Aegean Archaeology 6, 207-213. Oxford, Oxbow Books.

Godart, L. 1987. Le rôle du palais dans l'organisation militaire mycénienne. In E. Lévy (ed.), *Le Système Palatial en Orient, en Grèce et à Rome.* 237-253. Leiden, E.J. Brill.

Graziadio, G. 1991. The Process of Social Stratification at Mycenae in the Shaft Grave Period: A comparative examination of the evidence. *American Journal of Archaeology* 95(3), 403-440.

Greenhalgh, P. A. L. 1973. *Early Greek Warfare: Horsemen and chariots in the Homeric and Archaic ages.* Cambridge, Cambridge University Press.

Greenhalgh, P. A. L. 1980. The Dendra Charioteer. *Antiquity* 54, 201-205.

Haas, J. 1982. *The Evolution of the Prehistoric State.* New York NY, Columbia University Press.

Haggis, D. C. 1999. Some Problems in Defining Dark Age Society in the Aegean. In R. Laffineur and W.-D. Niemeier (eds.), *MELETEMATA: Studies in Aegean archaeology presented to Malcolm H. Wiener as he enters his 65th year.* Aegaeum 20, 303-308. Liège, Université de Liège.

Hall, J. M. 1995. Approaches to Ethnicity in the Early Iron Age of Greece. In N. Spencer (ed.), *Time, Tradition and Society in Greek Archaeology: Bridging the 'great divide'*. 6-17. London, Routledge.

Hall, J. M. 2007. *A History of the Archaic Greek World ca. 1200-479 BCE*. Malden MA, Blackwell.

Halstead, P. 1992a. Agriculture in the Bronze Age Aegean: Towards a model of palatial economy. In B. Wells (ed.), *Agriculture in Ancient Greece*. Skrifter utgivna av Svenska institutet i Athen 4°,XLII, 105-117. Stockholm, Svenska Institutet i Athen.

Halstead, P. 1992b. The Mycenaean Palatial Economy: Making the most of the gaps in the evidence. *Proceedings of the Cambridge Philological Society* 38, 57-83.

Hamilakis, Y. 2002. Too Many Chiefs? Factional competition in Neopalatial Crete. In J. Driessen, I. Schoep and R. Laffineur (eds.), *Monuments of Minos: Rethinking the Minoan palaces*. Aegaeum 23, 179-201. Liège, Université de Liège.

Harding, A. F. 2007. *Warriors and Weapons in Bronze Age Europe*. Archaeolingua Series Minor 25. Budapest, Archaeolingua Alapítvány.

Harrell, K. M. 2009. *Mycenaean Ways of War: The past, politics and personhood*. Unpublished PhD thesis, University of Sheffield.

Hitchcock, L. A. 2000. *Minoan Architecture: A contextual analysis*. Studies in Mediterranean Archaeology and Literature Pocket-Book 155. Jonsered, Paul Åströms Förlag.

Höckmann, O. 1980. Lanze und Speer im spätminoischen und mykenischen Griechenland. *Jahrbuch des Römisch-germanischen Zentralmuseums, Mainz* 27, 13-158.

Hood, M. S. F. 1954. Archaeology in Greece, 1954. *Archaeological Reports* 1, 3-19.

Hood, M. S. F. 1971. Late Bronze Age Destructions at Knossos. In A. Kaloueropoulou (ed.), *Acta of the 1st International Scientific Congress on the Volcano of Thera*. 377-383. Athens, Archaeological Services of Greece General Direction of Antiquities and Restoration.

Hood, M. S. F. 1973. The Eruption of Thera and its Effects in Crete in Late Minoan I In Πεπραγμένα του Γ' Διεθνούς Κρητολογικού Συνεδρίου. Vol. I. 111-118. Athens.

Hood, M. S. F. 1985. Warlike Destruction in Crete c.1450 BC. In Πεπραγμένα του Ε' Διεθνούς Κρητολογικού Συνεδρίου. Vol. I. 170-178. Iraklion.

Hope Simpson, R. and Hagel, D. K. 2006. *Mycenaean Fortifications, Highways, Dams and Canals*. Studies in Medterranean Archaeology 133. Sävedalen, Paul Åströms Förlag.

Iakovidis, S. 1983. *Late Helladic Citadels on Mainland Greece*. Monumenta Graeca et Romana IV. Leiden, E.J. Brill.

Immerwahr, S. A. 1990. *Aegean Painting in the Bronze Age*. University Park PA, Pennsylvania State University Press.

Kaniewski, D., van Campo, E., van Lerberghe, K., Boiy, T., Vansteenhuyse, K., Jans, G., Nys, K., Weiss, H., Morhange, C., Otto, T. and Bretschneider, J. 2011. The Sea Peoples, from Cuneiform Tablets to Carbon Dating. *PLoS ONE* 6(6), e20232. doi:20210.21371/journal.pone.0020232.

Keeley, L. H. 1996. *War Before Civilization: The myth of the peaceful savage*. Oxford, Oxford University Press.

Kelly, R. C. 2000. *Warless Societies and the Origin of War*. Ann Arbor MI, University of Michigan Press.

Kilian-Dirlmeier, I. 1993. *Die Schwerter in Griechenland (außerhalb der Peloponnes), Bulgarien und Albanien*. Prähistorische Bronzefunde IV,12. Stuttgart, Franz Steiner.

Kilian, K. 1987. L'Architecture des Résidences Mycéniennes: Origins et Extension d'une Structure du Pouvoir Politique Pendant l'Âge du Bronze Récent. In E. Lévy (ed.), *Le Système Palatial en Orient, en Grèce et à Rome*. 203-217. Leiden, E.J. Brill.

Kilian, K. 1988. The Emergence of Wanax Ideology in the Mycenaean Palaces. *Oxford Journal of Archaeology* 7(3), 291-302.

Knappett, C. 1999. Assessing a Polity in Protopalatial Crete: The Malia-Lasithi state. *American Journal of Archaeology* 103(4), 615-639.

Krzyszkowska, O. 1997. Cult and Craft: Ivories from the Citadel House area, Mycenae. In R. Laffineur and P. P. Betancourt (eds.), *TEXNH: Craftsmen, craftswomen and craftsmanship in the Aegean Bronze Age*. Vol. I. Aegaeum 16, 145-150. Liège, Université de Liège.

Laffineur, R. (ed.). 1999. *POLEMOS: Le contexte guerrier en Égée à l'âge du Bronze*. 2 Vols. Aegaeum 19. Liège, Université de Liège.

LeBlanc, S. A. 2003. *Constant Battles: The myth of the peaceful, noble savage*. New York NY, St. Martin's Press.

Lejeune, M. 1972. La civilisation mycénienne et la guerre. In M. Lejeune, *Mémoires de philologie mycénienne. Troisième Série (1964-1968)*. Incunabula Graeca 43, 57-77. Rome, Edizioni dell'Ateno.

Littauer, M. A. and Crouwel, J. H. 1983. Chariots in Late Bronze Age Greece. *Antiquity* 57, 187-192.

Littauer, M. A. and Crouwel, J. H. 1996. Robert Drews and the Role of Chariotry in Bronze Age Greece. *Oxford Journal of Archaeology* 15(3), 297-305.

Malkin, I. 2011. *A Small Greek World: Networks in the ancient Mediterranean*. Oxford, Oxford University Press.

Mann, M. 1986. *The Sources of Social Power: A history of power from the beginning to AD 1760*. Vol. I. Cambridge, Cambridge University Press.

Maran, J. 2001. Political and Religious Aspects of Architectural Change on the Upper Citadel of Tiryns. The case of Building T. In R. Laffineur and R. Hägg (eds.), *POTINA: Deities and Religion in the Aegean Bronze Age*. Aegaeum 22, 113-123. Liège, Université de Liège.

Maran, J. 2004. The Spreading of Objects and Ideas in the Late Bronze Age Eastern Mediterranean: Two case examples from the Argolid of the 13th and 12th centuries BC. *Bulletin of the American Schools of Oriental Research* 336, 11-30.

Maran, J. 2006. Coming to Terms with the Past: Ideology and power in Late Helladic IIIC. In S. Deger-Jalkotzy and I. S. Lemos (eds.), *Ancient Greece: From the Mycenaean palaces to the age of Homer*. Edinburgh Leventis Studies 3, 123-149. Edinburgh, Edinburgh University Press.

Mee, C. and Cavanagh, W. G. 1984. Mycenaean Tombs as Evidence for Social and Political Organisation. *Oxford Journal of Archaeology* 3(3), 45-65.

Mellersh, H. E. L. 1970. *The Destruction of Knossos: The rise and fall of Minoan Crete*. London, Hamish Hamilton.

Middleton, G. D. 2010. *The Collapse of Palatial Society in LBA Greece and the Postpalatial Period*. British Archaeological Reports International Series 2110. Oxford, BAR Publishing.

Miller, D. 1989. The Limits of Dominance. In D. Miller, M. Rowlands and C. Tilley (eds.), *Domination and Resistance*. 63-79. London, Unwin Hyman.

Molloy, B. P. C. 2006. *The Role of Combat Weaponry in Bronze Age Societies: The cases of the Aegean and Ireland in the Middle and Late Bronze Age*. Unpublished PhD thesis, University College Dublin.

Molloy, B. P. C. 2008. Martial Arts and Materiality: A combat archaeology perspective on Aegean swords of the fifteenth and fourteenth centuries BC. *World Archaeology* 40(1), 116-134.

Molloy, B. P. C. 2009. For Gods or men? A reappraisal of the function of European Bronze Age shields. *Antiquity* 83, 1052-1064.

Molloy, B. P. C. 2010. Swords and Swordsmanship in the Aegean Bronze Age. *American Journal of Archaeology* 114, 403-428.

Murray, O. 1993. *Early Greece* (2nd ed.). London, Fontana.

Nafplioti, A. 2008. 'Mycenaean' political domination of Knossos following the Late Minoan IB destructions on Crete: negative evidence from strontium isotope ratio analysis ($^{87}Sr/^{86}Sr$). *Journal of Archaeological Science* 35, 2307-2317.

Nakassis, D., Galaty, M. L. and Parkinson, W. A. 2010. State and Society. In E. H. Cline (ed.), *The Oxford Handbook of the Bronze Age Aegean (ca. 3000-1000 BC)*. 239-250. Oxford, Oxford University Press.

Nelson, M. C. 2001. *The Architecture of Epano Englianos, Greece*. Unpublished PhD thesis, University of Toronto.

Niemeier, W.-D. 1995. Aegina – First Aegean 'State' Outside of Crete? In R. Laffineur and W.-D. Niemeier (eds.), *POLITEIA: Society and state in the Aegean Bronze Age*. Aegaeum 12, 73-80. Liège, Université de Liège.

Nordquist, G. C. 1987. *A Middle Helladic Village: Asine in the Argolid* Acta Universitatis Upsaliensis Boreas 16. Uppsala, Academia Upsaliensis.

Otterbein, K. F. 1989. *The Evolution of War: A cross-cultural study* (3rd ed.). New Haven CT, HRAF Press.

Otterbein, K. F. 2004. *How War Began*. Texas A&M University Anthropology Series 10. College Station TX, Texas A&M University Press.

Palaima, T. G. 1999. Mycenaean Militarism from a Textual Perspective. Onomastics in context: lāwos, dāmos, klewos. In R. Laffineur (ed.), *POLEMOS: Le contexte guerrier en Égée à l'âge du Bronze*. Vol. II. Aegaeum 19, 367-379. Liège, Université de Liège.

Palaima, T. G. 2006. Wanaks and Related Power Terms in Mycenaean and Later Greek. In S. Deger-Jalkotzy and I. S. Lemos (eds.), *Ancient Greece: From the Mycenaean palaces to the age of Homer*. Edinburgh Leventis Studies 3, 53-71. Edinburgh, Edinburgh University Press.

Palmer, L. R. 1977. War and Society in a Mycenaean Kingdom. In *Armées et Fiscalité dans le Monde Antique*. Colloques Nationaux du Centre National de le Recherche Scientifique 936, 35-64. Paris, Éditions du Centre National de le Recherche Scientifique.

Papadopoulos, A. 2006. *The Iconography of Warfare in the Bronze Age Aegean*. Unpublished PhD thesis, University of Liverpool.

Papadopoulos, T. J. 1979. *Mycenaean Achaea*. Vol. I. Studies in Mediterranean Archaeology 55. Gothenburg, Paul Åströms Förlag.

Papadopoulos, T. J. 1998. *The Late Bronze Age Daggers of the Aegean I: The Greek mainland*. Prähistorische Bronzefunde VI,11. Stuttgart, Franz Steiner.

Parker Pearson, M. 2005. Warfare, Violence and Slavery in Later Prehistory: An introduction. In M. Parker Pearson and I. J. N. Thorpe (eds.), *Warfare, Violence and Slavery in Prehistory*. British Archaeological Reports Int. Ser. 1374, 19-35. Oxford, BAR Publishing.

Parkinson, W. A. and Galaty, M. L. 2007. Secondary States in Perspective: An integrated approach to state formation in the prehistoric Aegean. *American Anthropologist* 109(1), 113-129.

Pinker, S. 2012. *The Better Angels of Our Nature: A history of violence and humanity*. London, Penguin.

Rehak, P. 1995. Enthroned Figures in Aegean Art and the Fuction of the Mycenaean Megaron. In P. Rehak (ed.), *The Role of the Ruler in the Prehistoric Aegean*. Aegaeum 11, 95-118. Liège, Université de Liège.

Renfrew, C. 1972 [2011]. *The Emergence of Civilisation: The Cyclades and the Aegean in the Third Millennium BC* (Reprint ed.). Oxford, Oxbow Books.

Renfrew, C. 1975. Trade as Action at a Distance: Questions of integration and communication. In J. A. Sabloff and C. C. Lamberg-Karlovsky (eds.), *Ancient Civilization and Trade*. 3-59. Albuquerque NM, University of New Mexico Press.

Richardson, S. F. C. 2012. Early Mesopotamia: The presumptive state. *Past and Present* 215, 3-49.

Sahlins, M. D. 1963. Poor Man, Rich Man, Big-Man, Chief: Political types in Melanesia and Polynesia. *Comparative Studies in Society and History* 5(3), 285-303.

Sahlins, M. D. and Service, E. R. (eds.). 1960. *Evolution and Culture*. Ann Arbour MI, University of Michigan Press.

Sandars, N. K. 1961. The First Aegean Swords and Their Ancestry. *American Journal of Archaeology* 65, 17-29.

Sandars, N. K. 1963. Later Aegean Bronze Swords. *American Journal of Archaeology* 67, 117-153.

Sandars, N. K. 1964. The Last Mycenaeans and the European Late Bronze Age. *Antiquity* 38, 258-262.

Sandars, N. K. 1978. *The Sea Peoples: Warriors of the ancient Mediterranean 1250-1150 BC*. London, Thames and Hudson.

Schliemann, H. 1878. *Mycenæ : a narrative of researches and discoveries at Mycenæ and Tiryns*. London, J. Murray.

Schoep, I. 2002. Social and Political Organization on Crete in the Proto-Palatial Period: The case of Middle Minoan II Malia. *Journal of Mediterranean Archaeology* 15(1), 101-132.

Schon, R. 2010. Think Locally, Act Globally: Mycenaean elites and the Late Bronze Age world-system. In W. A. Parkinson and M. L. Galaty (eds.), *Archaic State Interaction: The eastern Mediterranean in the Bronze Age*. 213-236. Santa Fe NM, School for Advanced Research Press.

Service, E. R. 1975. *Origins of the State and Civilization: The process of cultural evolution*. New York NY, W.W. Norton.

Shelmerdine, C. W. 1996. From Mycenae to Homer: The next generation. In E. De Miro, L. Godart and A. Sacconi (eds.), *Atti e memorie del secondo Congresso internazionale di micenologia*. Vol. I. Incunabula Graeca 98, 467-492. Rome, Gruppo editoriale internazionale.

Shelmerdine, C. W. 2001. Review VI: The Palatial Bronze Age of the Southern and Central Greek Mainland. In T. Cullen (ed.), *Aegean Prehistory: A review*. American Journal of Archaeology Supplement 1, 329-383. Boston MA, Archaeological Institute of America.

Shelmerdine, C. W. 2006. Mycenaean Palatial Administration. In S. Deger-Jalkotzy and I. S. Lemos (eds.), *Ancient Greece: From the Mycenaean palaces to the age of Homer*. Edinburgh Leventis Studies 3, 73-87. Edinburgh, Edinburgh University Press.

Shelmerdine, C. W. 2008. Mycenaean Society. In Y. Duhoux and A. Morpurgo Davies (eds.), *A Companion to Linear B: Mycenaean Greek texts and their world*. Vol. I. Bibliothèque des Cahiers de l'Institut de Linguistique de Louvain 120, 115-158. Louvain-la-Neuve, Peeters.

Shelmerdine, C. W. and Bennet, J. 2008. Economy and Administration. In C. W. Shelmerdine (ed.), *The Cambridge Companion to the Aegean Bronze Age*. 289-309. Cambridge, Cambridge University Press.

Sherratt, A. G. and Sherratt, E. S. 1991. From Luxuries to Commodities: The nature of Mediterranean Bronze Age trading systems. In N. H. Gale (ed.), *Bronze Age Trade in the Mediterranean*. Studies in Mediterranean Archaeology 90, 69-83. Jonsered, Paul Åströms Förlag.

Sherratt, E. S. 2006. LH IIIC Lefkandi: An overview. In D. Evely (ed.), *Lefkandi IV: The Late Helladic IIIC Settlement at Xeropolis*. British School at Athens Supplementary Volume 39, 303-309. London, British School at Athens.

Small, D. B. 1998. Surviving the Collapse: The oikos and structural continuity between Late Bronze Age and later Greece. In S. Gitin, A. Mazar and E. Stern (eds.), *Mediterranean Peoples in Transition: Thirteenth to early tenth centuries BCE*. 283-291. Jerusalem, Israel Exploration Society.

Small, D. B. 1999. Mycenaean Polities: States or estates? In M. L. Galaty and W. A. Parkinson (eds.), *Rethinking Mycenaean Palaces: New interpretations of an old idea*. Cotsen Institute of Archaeology at UCLA Monographs 41, 43-47. Los Angeles CA, Cotsen Institute of Archaeology, University of California.

Snodgrass, A. M. 1964. *Early Greek Armour and Weapons: From the end of the Bronze Age to 600 BC*. Edinburgh, Edinburgh University Press.

Snodgrass, A. M. 1971 [2000]. *The Dark Age of Greece: An archaeological survey of the eleventh to the eighth centuries BC* (Reprint ed.). Edinburgh, Edinburgh University Press.

Southall, A. 1999. The Segmentary State and the Ritual Phase in Political Economy. In S. K. McIntosh (ed.), *Beyond Chiefdoms: Pathways to complexity in Africa*. 31-38. Cambridge, Cambridge University Press.

Spyropoulos, T. G. 1974. Το ανακτορον του Μινυου εις τον Βοιωτικον Ορχομενον. *Αρχαιολογικά ανάλεκτα εξ Αθηνών* 8(3), 313-325.

Spyropoulos, T. G. 1998. Pellana, The Administrative Centre of Prehistoric Laconia. In W. G. Cavanagh and S. E. C. Walker (eds.), *Sparta in Laconia: The archaeology of a city and its countryside*. British School at Athens Studies 4, 28-38. London, British School at Athens.

Stager, L. E. 1998. The Impact of the Sea Peoples in Canaan (1185-1050 BCE). In T. E. Levy (ed.), *The Archaeology of Society in the Holy Land*. 332-348. London, Leicester University Press.

Starr, C. G. 1977. *The Economic and Social Growth of Early Greece*. New York NY, Oxford University Press.

Tartaron, T. F. 2008. Aegean Prehistory as World Archaeology: Recent trends in the archaeology of Bronze Age Greece. *Journal of Archaeological Research* 16, 83-161.

Thorpe, I. J. N. 2005. The Ancient Origins of Warfare and Violence. In M. Parker Pearson and I. J. N. Thorpe (eds.), *Warfare, Violence and Slavery in Prehistory*. British Archaeological Reports International Series 1374, 1-19. Oxford, BAR Publishing.

Turney-High, H. H. 1949. *Primitive War: Its practice and concepts*. Columbia SC, University of South Carolina Press.

Underhill, A. 2006. Warfare and the Development of States in China. In E. N. Arkush and M. W. Allen (eds.), *The Archaeology of Warfare: Prehistories of raiding and conquest*. 253-285. Gainesville FL, University Press of Florida.

van de Moortel, A. 2009. The Late Helladic III C - Protogeometric Transition at Mitrou, East Lokris. In S. Deger-Jalkotzy and A. E. Bächle (eds.), *LH III C Chronology and Synchronisms III: LH III C Late and the transition to the Early Iron Age*. Veröffentlichungen der Mykenischen Kommission 30, 359-372. Vienna, Verlag der Österreichischen Akademie der Wissenschaften.

van Wees, H. 2004. *Greek Warfare: Myths and Realities*. London, Duckworth.

Vandkilde, H. 2003. Commemorative Tales: Archaeological responses to modern myth, politics, and war. *World Archaeology* 35(1), 126-144.

Vencl, S. 1984. War and Warfare in Archaeology. *Journal of Anthropological Archaeology* 3(2), 116-133.

Ventris, M. and Chadwick, J. 1973. *Documents in Mycenaean Greek* (2nd ed.). London, Cambridge University Press.

Vermeule, E. T. 1964. *Greece in the Bronze Age*. Chicago IL, University of Chicago Press.

Vermeule, E. T. and Karageorghis, V. 1982. *Mycenaean Pictorial Vase Painting*. Cambridge MA, Harvard University Press.

Voutsaki, S. 1995. Social and Political Processes in the Mycenaean Argolid: The evidence from mortuary practices. In R. Laffineur and W.-D. Niemeier (eds.), *POLITEIA: Society and state in the Aegean Bronze Age*. Aegaeum 12, 55-66. Liège, Université de Liège.

Voutsaki, S. 1999. Mortuary Display, Prestige and Identity in the Shaft Grave Era. In *Eliten in der Bronzezeit: Ergebnisse zweier Kolloquien in Mainz und Athen* Römisch-Germanisches Zentralmuseum Mainz Monographien 43,1, 103-117. Mainz, Verlag des Römisch-Germanisches Zentralmuseums.

Voutsaki, S., Triantaphyllou, S., Ingvarsson-Sundström, A., Sarri, K., Richards, M., Nijboer, A., Kouidou-Andreou, S., Kovatsi, L., Nikou, D. and Milka, E. 2007. Project on the Middle Helladic Argolid: A report on the 2006 season. *Pharos* 14, 59-99.

Walberg, G. 1995. The Midea Megaron and Changes in Mycenaean Ideology. *Aegean Archaeology* 2, 87-91.

Warren, P. M. 1991. The Minoan Civilisation of Crete and the Volcano of Thera. *Journal of the Ancient Chronology Forum* 4, 29-39.

Warren, P. M. 2001. Review of 'The Troubled Island: Minoan Crete before and after the Santorini eruption'. *American Journal of Archaeology* 105(1), 115-118.

Webb, M. C. 1975. The Flag Follows Trade: An essay on the neccessary interaction of military and commercial factors in state formation. In J. A. Sabloff and C. C. Lamberg-Karlovsky (eds.), *Ancient Civilization and Trade*. 155-209. Albuquerque NM, University of New Mexico Press.

Weber, M. 1947. *The Theory of Social and Economic Organization* (Translated by A. R. Henderson and T. Parsons). London, William Hodge.

Webster, D. 1975. Warfare and the Evolution of the State: A reconsideration. *American Antiquity* 40(4), 464-470.

Whitley, J. 1991. *Style and Society in Dark Age Greece: The changing face of a pre-literate society 1100-700 BC*. Cambridge, Cambridge University Press.

Whitley, J. 2002. Objects With Attitude: Biographical facts and fallacies in the study of Late Bronze Age and Early Iron Age warrior graves. *Cambridge Archaeological Journal* 12, 217-232.

Whitley, J. 2004-2005. Archaeology in Greece 2004-2005. *Archaeological Reports* 51, 1-118.

Wilkie, N. C. and Dickinson, O. T. P. K. 1992. The MME Tholos Tomb. In W. A. McDonald and N. C. Wilkie (eds.), *The Excavations at Nichoria in Southwest Greece Vol. II: The Bronze Age occupation*. 231-344. Minneapolis MN, University of Minnesota Press.

Wolpert, A. D. 2004. Getting Past Consumption and Competition: Legitimacy and consensus in the Shaft Graves. In J. C. Barrett and P. Halstead (eds.), *The Emergence of Civilisation Revisited*. Sheffield Studies in Aegean Archaeology 6, 127-144. Oxford, Oxbow Books.

Wright, J. C. 1995. From Chief to King in Mycenaean Society. In P. Rehak (ed.), *The Role of the Ruler in the Prehistoric Aegean*. Aegaeum 11, 63-81. Liège, Université de Liège.

Wright, J. C. 2004. The Emergence of Leadership and the Rise of Civilisation in the Aegean. In J. C. Barrett and P. Halstead (eds.), *The Emergence of Civilisation Revisited*. Sheffield Studies in Aegean Archaeology 6, 64-90. Oxford, Oxbow Books.

Wright, J. C. 2006. The Formation of the Mycenaean Palace. In S. Deger-Jalkotzy and I. S. Lemos (eds.), *Ancient Greece: From the Mycenaean palaces to the age of Homer*. Edinburgh Leventis Studies 3, 7-52. Edinburgh, Edinburgh University Press.

Wright, J. C. 2008. Early Mycenaean Greece. In C. W. Shelmerdine (ed.), *The Cambridge Companion to the Aegean Bronze Age*. 230-257. Cambridge, Cambridge University Press.

Yasur-Landau, A. 2010. *The Philistines and Aegean Migration at the End of the Late Bronze Age*. Cambridge, Cambridge University Press.

Soldiers and Civilians – A New Look at Asymmetric Warfare in the Eastern Roman Empire in the 1st to 3rd century AD

Birgitta Hoffman

Abstract

In our common portrayal of the Roman Imperial army the focus tends to be in the engagement between two armed regular forces, i.e. armies that are recognizable as that. However, during the course of its history the army also engaged with what today would be called insurgents, civilians or irregular militias, especially during times of civil war or in the aftermath of invasions. In the East and Africa these frequently appear to be drawn from the city population rather than the countryside, and the retribution for their action was frequently bourne by the community as a whole, leading in the past to models identifying the alienation of the members of the Roman army from their population. This paper will look at the evidence again and assess whether more integrative models based on shared perception of the 'pax Romana' may not be more useful in this context.

Introduction

In AD 238 the *Legio III Augusta* with its legate Capelianus defeated outside Carthage the massed militia troops of the usurpers Gordian I and II. Following the defeat large scale retributions for the supporters followed throughout Africa. The episode is described in Herodian (Herodian, 7.9.10):

....On his entry into Carthage, Capelianus massacred any prominent person, who had escaped the battle, and had no compution about robbing temples or confiscating private and public funds. He also attacked other cities that had destroyed dedications to Maximinus, killing the leading citizens and driving the lower class out of the territory. Fields and villages were turned over to the soldiers for burning and plunder, on the grounds that this was punishment for their offences against Maximinus. (Translation Whittaker).

The Scriptores Historiae Augustae (*The Two Maximini* 19.3-5) describes the events as follows:

Tunc Capelianus victor pro Maximino omnes Gordiani mortui partium in Africa interemit atque proscripsit nec cuiquam perpercit, prorsus ut ex animo Maximini videretur haec facere. Civitates denique subdidit, fana diripuit, donaria militibus divisit, plebem et principes civitatum concidit.

('And forthwith Capelianus the victor, in the name of Maximinus slew and outlawed all of the dead Gordian's party in Africa, sparing none. Indeed, he seemed to perform these duties quite in Maximinus' own temper. He overthrew cities, ravaged shrines, divided gifts among his soldiers and slaughtered common folk and nobles in the cities.')

The two descriptions of the events after the battle are very close and it cannot be ruled out that the author of the Scriptores Historiae Augustae may have drawn on Herodian at this point. Herodian himself does not claim to be an eyewitness of the events, but he does describe the events as they happened during his own lifetime (Herodian, 1.2.5). While this does not exclude that there may be mistakes in the accounts, it does provide the ancient historian with the valuable tool of having the 'personal comment' of a contemporary on these matters, although Herodian's professional position within the Roman empire appears as unclear as his origins (e.g. Magie 1969, IX-XXV).

The personal view of a contemporary of these events is in this case of particular importance, as Herodian's disapproval of the events unfolding in Africa is clear, and has been picked up by various scholars, especially Shaw (1983). He links this disapproval with an unusual level of violence displayed by the army to their 'own' population. To account for the possibility of this level of 'savagery', he proposes a 'terrible level of estrangement of the body of the soldiery and the local populace' (Shaw 1983, 144). He describes the *Legio III Augusta* as 'recruited wholly African', but not Numidian (Shaw 1983, 148, 144), which he seems to mean that he saw them originating from the old Army bases (and thus of immigrant stock), rather than coming from the Numidian indigenous population.

Brent explains this perceived phenomenon by referring to the new structures of loyalty imposed by the army after joining, applying the concept of 'total institution' (1983, 148). This term, defined as 'a community socially and culturally isolated from the civilian population it oversaw' (Haynes 1999, 8), has since been taken up by scholars studying the role of violence/military within the Roman empire and adapted. The events of AD 238 in Africa, are usually quoted as the clearest indication of this at work. In view of the large amount of work done on the *III Augusta* it seems, however, appropriate to revisit the occasion and scrutinize Shaw's original argument.

The Origins of the Legio III Augusta

Shaw (1983) stresses two facts about the *III Augusta*: that in AD 238 it was wholly recruited in Africa, but not from the local Numidian population. The question of the local Numidian population has to be addressed first. It is undoubtedly true that amongst the recruits of the *III Augusta* there are no members of the local tribes who identify themselves as such through their indigenous names in the epigraphic record. This, however, is a problem with the recruitment of the legions generally. Traditionally, eligibility for the Roman legions depended on Roman citizenship and the ability to understand Latin (the command language). Thus in North Africa the origins of legionaries are dominated by towns of colonial and municipal status. From the existing information only 53 out of nearly 700 cases give origins, referring to civitas or other peregrine communities (Le Bohec 1989, 526). Roman citizenship is most easily expressed epigraphically by the use of a Latin name, originally in the form of the tria nomina, but by the early 3rd century AD the use of all three names was rapidly dropping out of use. It is also known (at least for a slightly earlier period) that in the case of exotic names the Roman army was not above renaming its recruits upon enrolment.

For the period of AD 238 we must assume that the local Numidian tribes had been given citizenship under the *Constitutio Antoniniana*. Given the continued sole use of Latin within the army, as well as a traditional preference in the *Legio III Augusta* for Latin cognomina,[1] there is little reason why we should expect a sudden flood of Numidian names in the epigraphic record. More likely candidates would be soldiers commemorating themselves as Aurelius Donatus or using similar Latin 'African' names, i.e. names that represent Latin translations of Numidian and Punic Names. Thus Le Bohec's listing of soldiers of the period of the Severans to Maximinus Thrax (1989, 314-332) (the period that should concern us here most) lists 33 Aurelii, including a Saturus, a Saturninus and an Optatus, which may be translations from Punic names. Further examples can be found amongst the higher ranks, including an Aurelius Cittinus (CIL III2564=18052), for which an African origin is proposed by Le Bohec. These Latinisations of African names are well attested throughout North Africa, especially when bilingual inscriptions are available. Instead of a categorical ruling out of Numidian recruits as proposed by Shaw, the statement would be best amended to: it cannot be ruled out that by the early 3rd century AD local Numidian families were joining the legion, while African contribution to auxiliaries is attested since the 1st century AD (Le Bohec 1989, 512-517).

What Shaw's list shows is that there was a very high amount of recruitment from the soldiers of the Numidian towns: Shaw (1983), following Forni (1953; 1974), lists 13 towns and settlements in Southern Numidia that provided recruits. It is true, especially in the case of Timgad and Diana Veteranorum, that the chances are high for recruits to have descended from veteran settlers, especially in view of the citizenship requirement noted above. While Shaw (1983) separates these towns from the Numidian hinterland and singles them out as different, however, the same is not done for the colonies and municipia elsewhere, some of which have their origin in the settlement of Caesarian and Augustan veterans. It must also be stressed that by the early 3rd century AD the Numidian settlements would have been in existence for about 100 years, or roughly four generations. Under these circumstances it is highly problematic to differentiate between the indigenous and the veteran groups, and not see soldiers recruited in Numidia as part of a local recruitment pattern. Furthermore, by AD 238 these towns would have been part of an independent province of Numidia for nearly 40 years, completely separate from the province of Africa. It is debateable how far the area had diverged beforehand. Fentress (1979, 124 *passim*) clearly believes in a form of administration that was widely independent of Carthage since the 1st century AD, while Shaw chooses to stress the links between the two provinces in the late 1st and 2nd centuries AD. Experiences of re-unification in Germany show that after 40 years of separation there remains an automatic generic sense of 'belonging together', even if family links cannot be assumed, but that the primary level of identification (e.g. in supporting national teams etc. or in stress situations) is likely to run with the 'new country' rather than some past unified whole. Should Numidia have been administered in effect as an autonomous region, as envisaged by Fentress, from much earlier on, then these effects are likely to have been even more pronounced. Africa is thus likely to have been seen as a neighbouring province with a common past, but not necessarily as 'us'.

Le Bohec's revised lists for the recruitment for the *III Augusta* are able to further refine the image. For the period from AD 193-238 he has 525 records of origin (Table 1).

With 40% of the attested recruitment coming from Africa itself, however, it might be noted in Shaw's defence that there is a very high likelihood of family links within the area.[2] It is impossible to establish how high these contacts and feelings were, especially given the geographical constraints: the majority of the legion operated well south of the pertica of Carthage and the legionary base at Lambaesis are c.400 km apart. In the absence of means of mass communication and rapid transport, it is unlikely that contact was close, although in view of the exchange

---

[1] Le Bohec (1989, 530) compares the 95% of Latin cognomina in the *Leg III Augusta*, with the much lower levels of 50% in Northern Britain. If this is corrected just for the names connected with the British legions (based on RIB), the relationship of Latin to non-Latin Cognomina rises to c.85:15%, but is still lower than the African example.

[2] It could be argued that the likelihood of your origin being recorded increases the further removed from your place of residence/burial you were born (cp. the Syrian at Hadrian's Wall for example), leading to overrepresentation of people from further away. Given the propensity at Lambaesis to record castris and local origins in large numbers, it is possible that the recording of origins was here handled similarly for everyone and that the 40% attested might reflect a true percentage.

of letters between soldiers in Italy and their families in Egypt, it should not be underestimated.

In those cases one would expect at least some of the soldiers to return to their homes after the end of their military service. It is striking that during the early 3rd century AD no veterans are recorded outside Lambaesis. There is, however, a strong tendency even at Lambaesis for veterans not to record their position on gravestones, but more frequently on communal altars, such as the altar to IOM F dedicated for the welfare of the emperor (CIL III 2626=18099). Under these circumstances it is impossible to decide what became of the veterans after leaving the legions. It is possible that they did not leave Numidia in large numbers, but it should not be ruled out that, once retired, their former rank in the military was considered unsuitable for public commemoration. A further question often raised is of whether links between the nobility of the African towns and the legions existed. In the absence of any incontrovertible evidence, Le Bohec searched for parallels between unusual family names and members of the army, but found no convincing evidence that these links existed (Le Bohec 1989, 522). There is thus little to suggest that the Legio III Augusta would have felt particularly strong ties to the province of Africa.

While Africa certainly provided a substantial number of recruits, it is, thus, also clear that the legion at that time operated from an independent province, which had time to develop identities different to those expressed in the province of Africa. In the early 3rd century AD Numidia and its settlements (both military and non-military) provided about 54% of the personnel of the legion and should therefore be assumed to dominate any questions of identity within the legion.

The Identity of the Numidian Troops

Establishing the identity of the troops is essential to understanding how far they might have identified with the people they were fighting in AD 238. Shaw (1983, 148) stresses the unifying and separating character of army life. We have seen above that his statement that the legion did not integrate with civil Numidian society is open to reinterpretation with regard to the rural population, while the urban population appears to have made a substantial contribution to the legion.

While there are no unequivocal statements regarding how the army saw themselves, other indicators must be sought. One of these is religion: Africa was dominated in large part by a strong support for Tanit and Baal, and their Romanised counterparts Saturn and Iuno Caelestis. This image is not reflected in the religious dedications of the *Legio III Augusta*, which are strongly reminiscent of the official Roman cult. Le Bohec points out contemporary choices on imperial coinage (Le Bohec 1989, 549-550), while the African deities appear to be underrepresented.

No similar studies were available to compare this scenario with the religious choices outside the military in Numidia. Should this difference in cult preferences reflect a division between the military and civilians within Numidia, then this would provide strong support for Shaw's theory of the military as a community separated from the civilians around it.

For the moment, differences are only apparent between the religion of the civilian province of Africa and the military (stationed in Numidia). Thus it cannot be ruled out that this is a geographical difference, reflecting a difference in religious views between Numidia and Africa, rather than a difference of religious views between the military and the civilian sphere.

Both Fentress (1979) and Le Bohec (1989, 579-580) stress the divergent character of southern Numidia in comparison to the three parts of the African province, which were later known as Zeugitana, Byzacena and Tripolitania. More recent surveys in both the Tell Tunisien (Peyras 1991) and the Segermes valleys support the differences in character between the provinces. Such differences in geography, settlement pattern and origin of population can create differences in identity, not only in self-definition, but often more importantly in the identity ascribed to the populace by outsiders.

At this point it is necessary to return to Herodian himself. As noted at the beginning of this paper, it is Herodian's disapproval that suggested to Shaw and Haynes that the incident of AD 238 is different from other military interventions. As outlined above, Shaw suggested that this was due to the fact that the local army turned on the civilians. We have seen that the arguments for describing the *Legio III Augusta* as 'local' are far from watertight. But regardless of our view of events, it is necessary to verify that Herodian perceived the *Legio III Augusta* as sharing the same identity as the people they attacked.

Herodian identifies the people who support (or rather start) the Gordian rebellion as Libues, or Libyans (e.g. 7.4.1), this being the common Greek term for Africans outside Egypt. His reference to Carthage in the same paragraph makes it clear that he is speaking of the province of Africa. Further references to Libya/Africa follow (7.4.4. and 7.5.8). In the latter he makes the equation clear, by stating: 'In addition to his own title they gave him the name of Africanus after themselves, the name given to the Libyans in the south by those who speak the language of the Romans.' (Translation Whittaker). In comparison the name given to Capelianus' army is linked to the Mauri: 'A senator called Capelianus was the commander of the part of Mauretania under Roman jurisdiction called Numidia.' (Translation Whittaker).

Later during the battle Capelianus' forces are described thus (Herodian, 7.9.6): 'The Numidians were crack spearmen and expert riders, able to control their horses at the gallop without reins and using only a riding crop.' (Translation Whittaker).³

At no point in his account does Herodian mention any army units by name, and the units described here are certainly not the *Legio III Augusta*, which was later punished. Herodian, therefore, does not see the two opponents as the same people: the army of Numidia linked with the Mauri, in whose territory they lived and whom they regularly fought, and on the other side the Africani/Libyans. Shaw's problem of having to explain how the army could turn on its own civilian population was thus not apparent to Herodian and cannot, therefore, be the reason for his disapproval.

How to Re-Establish Order Amongst Civilians in Herodian's View

This raises for the interpreter the problem of identifying the reason for the censure. To a modern reader, the idea of large scale violent retribution after what is in effect a civil uprising is disturbing, and it is tempting to suppose that this response may be behind Herodian's disapproval. A careful perusal of Herodian makes it clear, however, that this is far from the case, as AD 238 is only one of about a dozen incidents of a similar nature.

A very similar incident occurs during the Civil War of Septimius Severus vs. Pescennius Niger (Herodian, 3.3.3-5):

> ...There was an outbreak of local rivalry in Syria by Laodiceia which hated Antioch, and in Phoenicia by Tyre through enmity with Berytus. When both these cities heard that Niger had been routed they seized their chance to strip Niger of his honours and to recognise Severus. Niger heard the news when he reached Antioch, and though he had acted generously up to now, he was quite reasonably angered by this defiant revolt. So he dispatched against the two cities some Maurousian spearmen that he had with him and a section of archers with orders to kill any who met them, to seize the movable property in the cities and burn down the buildings. The Maurousii are extremely bloodthirsty and ready for any desperate act because of their complete disregard for death or personal danger. They fell upon Laodikeia without warning and subjected the city and its inhabitants to all kinds of outrage. They hurried on to Tyre and destroyed the whole city in flames after looting and killing. (Translation Whittaker).

This section rivals the Carthage episode in terms of violence, but Herodian points out that Niger is 'reasonably angered'. Although the Mauri are once again singled out for their 'effectiveness', this is not criticized but appears to be the rationale of their choice. As in the case of Carthage the troops are identified by Herodian as being different from the local population and unattached to them, as they had been with Niger up to then.

Niger is not the only one who is described using this method. After the battle of Issus and the death of Niger, Septimius is described as follows (Herodian, 3.4.7): 'Now that Niger was out of the way, Severus ruthlessly punished all Niger's partisans, regardless of whether they had joined him voluntarily or had been forced to do so....'

While this lacks the colourful details provided for Laodikeia and Carthage, it nevertheless suggests the same treatment, a treatment that the population of Antioch appears to have expected, as they evacuate the city after the lost battle (Herodian, 3.4.6.), and the missing detail is provided later (Herodian, 3.5.6):

> Severus' actions against Niger's generals had detracted from his reputation because, after putting pressure on them through their children to betray Niger (...), he made use of their services, but once they had achieved his aims he destroyed them and their children. It was these acts that really showed his underlying character. (Translation Whittaker).

The criticism here is apparently not that he moves against the supporters of the other side, but that he uses people who had left Niger and afterwards kills them, effectively going back on his word. He fits in well with Herodian's general depiction of Severus as a crafty trickster, rather than an 'honourable man'.

The destruction of a city that has supported the opposing side is again reported without criticism in the case of Byzantium (Herodian, 3.6.9) and Lyon (Herodian, 3.7.3-7), both of which are sacked by Pannonian troops, who again were foreigners to those areas.

After the fall of Albinus, the actions against his supporters are again those of execution and confiscation of property (Herodian, 3.8.2). In fact, these actions only incur the criticism of Herodian (3.8.7), when he suggests that the persecutions are motivated by greed, rather than revenge. We once again find that Herodian criticizes the motive, but not the action itself.

It has to be stressed that the tone of these events is very different from the massacres that are described by Herodian for Caracalla's reign, such as the massacre in Alexandria (Herodian, 4.9).

In comparison to these events the African incident of AD 238 stands out merely for its mention of the plundering of temple treasures. This, however, follows an earlier comment of Herodian concerning Maximinus Thrax (Herodian, 7.3.5), who allowed the melting down of temple treasures to raise money, a fact that did not endear him to the population.

Given the importance that Herodian placed on the underlying motives for the actions undertaken during the repression of both the Niger and Albinus episodes, it is in

---
³ This suggests that we are dealing with non-Roman troops, i.e. not the legion.

this context important to draw attention to the concluding sentence of the Carthage revolt (Herodian, 7.9.11):

> ...on the grounds that this was a punishment for the their offences against Maximinus, though in fact Capelianus was quietly canvassing the loyalty of the troops for himself. If anything were to take a wrong turn in Maximinus' fortunes, he intended to make a bid for the empire himself with the aid of a loyal force of soldiers. (Translation Whittaker).

The Scriptores Historiae Augustae's version (*The Two Maximini* 19.5), which admittedly may be a copy, puts it thus: *Ipse praeterea militum animos sibi conciliabit, proludens ad imperium, si Maximinus perisset.* ('At the same time he strove to win over the affections of his soldiers, playing for the imperial power himself in the event that Maximinus perished.' (Translation Magie)).

This suggests that Herodian once again objected to the underlying motive, a possible bid for power. Herodian did not see the methods employed as despicable or unusual. What mattered to him was the fact that this was not done to re-establish order, but to prepare a separate bid for power. In effect, Herodian suggests that Capelianus is undermining the Roman justice system by using the violence which is 'normal', and perhaps perceived as 'cleansing' for the state, to start another rebellion. The subject of Capelianus is not resumed at any point in the surviving narrative, although the turn of phrase suggests a 'cliff-hanger', rather than a final verdict on the general, which may imply that further texts are missing, although we will probably never be able to answer this particular textual problem.

Conclusion

This article began with Herodian's description of the end of the Gordian revolt in Africa in AD 238. Interpreting the events, Shaw argues that this episode suggests that the army in Africa, particularly the legion, may have been emotionally detached and isolated from the remainder of the civilian population. Reviewing the evidence, it is hard to see how the Gordian rebellion supports such a claim, as our surviving source makes it clear that the issue for him is not one of an army turning on its own citizens, but instead one in which the governor of a neighbouring province re-establishes order after a revolt. At no point in the historical narrative is there a suggestion that there is a conflict of identities, but the description speaks of Africani/Libyans vs. Numidians/Mauri. Herodian's remaining text reinforces the impression that he did not see this as a unique incident, as the reaction of the victor can be paralleled in other civil war/revolt scenarios which involve the civilian population. While this paper has only attempted to review the evidence for Herodian, there is no shortage of ruthlessness in the putting down of revolts or extinguishing the loosing side in a civil war. This begins in the 1st century BC with Marius' and Sulla's proscriptions, but also includes the Saturninus Rising and the events of the Year of the Four Emperors. Modern perceptions of war prefer to demand some distance between the victims and the perpetrators of these 'cleansings', and in the case of the examples named by Herodian, the troops involved are rarely stationed in the same area as their victims and are usually brought in from elsewhere. It should, however, be stressed that history teaches us that such 'separation' is not necessary to create such savagery. In the aftermath of the 1745 rising in Scotland very similar acts were conducted by the Hannoverian army, which consisted to a large part of Scots who had relatives on the other side, a fact frequently remarked upon even at the time (for an introduction see Craig 1997). It seems that under the exceptional circumstances of revolts/civil wars the human psyche unfortunately does not need recourse to such concepts as 'total institutions', although as recent work has shown they do have their place in studies of the peace-time Roman military.

Bibliography

The Scriptores Historiae Augusta (Translation and Introduction D. Magie). 1960. Loeb Classical Library. Harvard University Press.

Herodian (Translation and Introduction C. R. Whittaker). 1969. Loeb Classical Library. Harvard University Press.

Craig, M. 1997. *Damn Rebel Bitches. The Women of the '45*. Edinburgh and London, Mainstream Publishing.

Fentress, E. 1979. *Numidia and the Roman Army*. British Archaeological Reports International Series 53. Oxford, BAR Publishing.

Forni, G. 1953. *Il reclutamento delle legioni da Augusto a Diocleziano*. Milan-Rome: Fratelli Bocca.

Forni, G. 1974. Estrazione etnica e sociale delle legioni nei primi tre secolo d'impero. *Aufstieg und Niedergang der römischen Welt* 2,1, 339-391.

Haynes. I. 1999. Introduction: The Roman Army as a Community. In A. Goldsworthy and I. Haynes (eds.), *The Roman Army as a Community*, 7-15. Journal of Roman Archaeology supplementary Series 34. Portsmouth, RI, Journal of Roman Archaeology.

Le Bohec, Y. 1989. *La troisième Légion Auguste*. Paris, CNRS.

Peyras, J. 1991. *Le tell Nord-est Tunisien dans l'antiquité*. Paris, CNRS.

Shaw, B. E. 1983. Soldiers and Society: The army in Numidia. *OPUS* 2, 133-157.

Origin	Number (Total 525)	Percentage of Total	
Foreigners	24	4%	4%
Northern provincia Africa	120	23%	40%
Southern provincia Africa	83, of which 36 come from the military colonies of Ammaedara and Theveste.	16%	
Tripolitania	6	1%	
Northern Numidia	56	10%	55%
Southern Numidia	48	9%	
Castris	188	36%	

Table 1: Recruitment of the *Legio III Augusta* in AD 193-238. After Le Bohec (1989, 494-503).

Militarization, or the Rise of a Distinct Military Culture? The East Roman Ruling Elite in the 6th Century AD

Conor Whately

Abstract

When discussing the aristocracy of the barbarian kingdoms of the post-Roman West, it is common for modern scholars to remark on their militarization, a process which is said to have started as a result of the separation of the civilian and military elite in the 3rd century AD, and the barbarization that came with Rome's increasing contact with its neighbours. This is in marked contrast to the East Roman Empire, whose aristocracy scholars such as Chris Wickham and Walter Goffart have said was not militarized before the middle of the 7th century AD. This is surprising, especially considering that it was the East that was able to endure, while the West fragmented; the military character of the eastern empire surely contributed to its political survival. In this short paper I look at textual and iconographic evidence, including Procopius' Gothic Wars *and the* Prosopography of the Later Roman Empire (PLRE), *which suggests that at least certain components of the aristocracy of the East were also culturally and socially militarized, and that this process of militarization started earlier than has generally been accepted, in the 6th century AD.*

Introduction

Militarization, the gearing of a state for war, is an all-encompassing process that affects everything from foreign policy to social structuring. The term has been widely applied to various periods in Roman history. During the Republic, in light of its various expansionist activities, Rome is said to have been militarized, and generally quite aggressive. After the demilitarization of the Roman citizenry during the Principate, this process is said to have gained momentum in the 3rd century AD and continued into the 4th and 5th centuries AD as the West fractured into a number of smaller kingdoms. In the East, however, we have a different story, for the East Roman state is usually not considered to have attained the level of militarization achieved by its western counterparts before the AD 650s.

In this preliminary study I want to ask whether we can find signs of this process beginning much earlier in the 6th century AD, or if what evidence we have suggests something else, namely the rise of a military elite. The former applies to society as a whole, while the latter concerns one social group, whose success is based on warfare, an important difference in assessing the scale of any major historical change. Another possibility is that this is not an either/or question, as a military elite could very well arise in a militarized society. To address this question I am going to look at four issues: the definitions of 'militarization', 'aristocracy', 'elite', and 'Roman'; highlight select characteristics of East Rome's principal neighbours; look at war, imperial propaganda, and the elite imagination; and lastly look at the careers of the elite. The main question I want to address through all this is whether there was a militarization of the elite in general, or rather a rise of a separate military elite as part of the wider army community.

Definitions: 'Militarization', 'Elite', 'Aristocracy', and 'Roman'

Militarization

In attempting to define this term I looked at a very select body of texts out of necessity. Among them, no overall consensus prevails on what constitutes militarization, and some fail to define what they mean by militarization.

Among students of the Roman army a varied assortment of views is found. MacMullen, for example, in his *Soldier and Civilian in the Later Roman Empire* (1963), while claiming that the late empire was in some sense militarized, notes that 'the emperor…drew close to his troops', and that 'the balance of power and prestige inclined…towards army officers' (MacMullen 1963, 176). His focus rests squarely on governmental changes (MacMullen 1963, 65), rather than societal ones. Shaw (1999, 146), like a number of scholars before him (Cornell 1993; Kuefler 2001, 45), speaks of a demilitarized centre and a militarized periphery during the Principate. Here he is mostly concerned with physical structures and the presence, or absence, of Roman armies. Like Macmullen, Eich (2007, 511-515) uses militarization in regard to the changes in government in the 3rd century AD, particularly career changes (soldier versus civilian), though he also notes the preponderance of conflict.

The main problems with these definitions are that they are too vague or too limiting. A better one comes from Wilson (2008), whose primary focus is not on the Roman world. In an important study he focuses on defining military culture using examples from early modern and modern history, while also managing to put military culture in relation to militarization.

Wilson (2008, 11) suggests that military culture is a form of institutional culture, and so considers the army as an institution; the same claim – army as institution – has previously been made about the imperial Roman army (Pollard 1996; Potter 2004, 132). While elaborating this notion of the army as an institution Wilson notes five aspects: its mission, its relationship to the state and other institutions, its relationship to society, its internal structure, and its access to resources (money, technology, and education). Militarization is discussed at the end, in which a distinction is made between 'militarization', which he defines as the capacity of a state to wage war, and 'militarism', the mental and cultural willingness to embark on it, (cf. Hunt 2010, 226-231) with a militarized state being one that is materially organized to wage war (Wilson 2008, 40).

Wilson (2008, 40) then relates his 'discussion of military culture to the broader debate on the wider presence and influence of military attitudes and values. Militarisation has political, social, economic, and cultural dimensions.' These various dimensions he defines as follows: political militarization is 'the extent to which the state structure is geared for war'; societal militarization is 'the proportion of the population incorporated into military institutions, and, by extension, involved in other preparations for war, such as working in arms industries'; economic militarization is 'a matter of resource mobilisation'; and cultural militarization 'entails the wider presence of military culture in society beyond military institutions' (Wilson 2008, 40-41). Bearing these definitions in mind, as well as the question which frames this paper, 'militarization of the elite' can be equated with Wilson's societal and cultural militarization, while the rise of a military culture can be equated with his military culture. On the other hand, they are not mutually exclusive ideas.

Aristocracy and Elite

The next term which must be defined is the particular group within East Roman society which forms the focus of this paper: are they an aristocracy or an elite, and how are these terms relevant? Both terms are, in fact, problematic (Börm 2010, 164).

Aristocracy is a term used by Harris in his *War and Imperialism in Republican Rome* (1979) to describe the leading members of the state, by Lendon in *Empire of Honour* (1997) as part of a study of the imperial Roman government, and by Salzman in *The Making of a Christian Aristocracy* (2002), which looks at the conversion of the western senatorial aristocracy in the 4th and 5th centuries AD. All three use it to refer to a subgroup of the rich land-owning group within Roman society – 'a group defined by its shared values, and in particular by its members' esteem of the same qualities' (Lendon 1997, 37; cf. Salzman 2002, 24). However, aristocracy often denotes a privileged group of interconnected families, and while this is appropriate for much of Roman history, it is less so for the period under review.

An alternative term is 'elite'. It has been used frequently by scholars like Mattern (1999) in her study of Roman ideology and foreign policy, by authors in Haldon and Conrad (2004) on elites in the late antique and Islamic near east, and by Börm (2010) in his paper on select aspects of the relationship between rulers and the elite in late antique Rome and Iran. A potential objection to the term 'elite' is that there could be multiple levels of elites, and overlap within those levels. However, as long as one makes clear who is being included in a particular category, it matters little. As such, the term 'elite' will be used here to denote the 'ruling elite'. Here I follow the definition of Haldon (2004), for whom the ruling elite represents the uppermost group in society. Moreover, by using the term elite I obviate both the problems of familial associations, regardless of whether it may have been an important component, and the potentially different backgrounds of this group, especially as regards geographic origins.

Roman

I must also clarify what I mean by 'Roman'. For if the elites whom I am focusing on are not socially or culturally militarized, or are part of a different culture, this will change the nature of the investigation.

As with the Roman imperial armies and their component parts of *auxilia* and *numeri*, the armed forces of the 6th century AD were heterogeneous. Yet, while questions of Romanity and barbarization are rarely raised about the imperial period, they often feature in discussions of late antiquity. In discussions of the organization of the 6th century AD armed forces, and in the texts from which that information hails, references occur for all kinds of troops: Armenians, Goths, Huns, Iberians, Isaurians, and Thracians. That these names are used primarily by classicizing historians such as Procopius and Agathias has led some scholars to think that the problem of barbarization was something these historians were highlighting. A careful reading of the material, however, suggests otherwise, and instead points towards these historians' use of standard historiographical *topoi*, as noted by numerous scholars (Müller 1912, 101-114; Grosse 1920, 272-294; Teall 1965; Whitby 1995, 103-110; Kaldellis 2007, 209-217; and Rance 2005, 443-447).

Leading generals of the era are often identified by the region from which they originated. Thus Amory (1997, 277-313) spends much time elaborating what he calls 'Balkan Military Culture', Croke (2001, 88-101) discusses the Illyrian generals resident in Constantinople, and Goffart (2006, 187-229) equates barbarization with social militarization, or the rise of a multi-ethnic military aristocracy. There is therefore a question as to whether these are a group of military officials different from those with which this paper is concerned. Ought these commanders from the Balkans serving Rome, and ultimately based in Constantinople, be considered as a distinct group from the Romans? My answer is no. Although it is important to recognize the differences between late antiquity and the present, these Illyrians serving Rome may be thought of as akin to Quebeckers in the 21st century AD serving Canada. This does not, of course, obviate the existence of identifiable groups serving the East Roman Empire who were not Roman, such as Persian or Vandal prisoners of war. They are, however, likely to have served under what may be termed Roman commanders, and though the identity of certain troops might be non-Roman, those in charge of the armed forces invariably were.

East Rome's Neighbours and their Militarized Elites

At the beginning of the 6th century AD the East Roman empire was surrounded by a number of states of varying size. To the East was Sasanid Persia, Rome's great political rival, and the only state to which it accorded equal status (Peter the Patrician, fragment 13; Theophylact Simocatta, 4.11.2; Mitchell 2007, 389; Canepa 2009). To the north were a host of peoples such as the Avars, Gepids, Heruls, Huns, and Slavs, perhaps best characterized as tribal groups rather than states, and best classed as minor powers, though they often caused considerable problems to East Rome, particularly when the latter was engaged in other conflicts. To the southwest lay Vandal Africa, to the west Ostrogothic Italy, and to the northwest Merovingian Gaul. During the course of the 6th century AD these states were often at war with one another, and with East Rome, and so, despite the presumed dissolution of Mediterranean unity, there continued to be considerable interaction.

In modern discussions, the elites of these various kingdoms and tribal groups are invariably characterized as militaristic. Among barbarian kingdoms in the post-Roman west Halsall (2003, 30) notes that 'throughout this period aristocrats were expected to be warriors'. The Vandals are considered to have had a militarized ruling class (Wickham 2005, 91; Halsall 2007, 296). The Ostrogoths are also said to have had a military elite with a warrior ethos (Heather 1996, 322-326; Whitby 2000b, 472; Kouroumali 2005, 170). Amory (1997, 77) notes that the Ostrogothic king Wittigis, the first king to face the East Romans during Belisarius' campaigns, was 'chosen king amidst a circle of spears, on a shield, in the open field'. With respect to the Ostrogoths' northern neighbours the Franks, Wickham (2005, 203) notes: 'Francia can indeed in part serve as a model, in particular for the militarization of aristocratic identity'. Halsall (2007, 303) modifies this claim and suggests that the ruling elite grew out of the Roman army. Van Dam (2005, 205 ff.) holds comparable views. Finally, for Howard-Johnston (2008, 125; 2010), the Sasanid Persians are 'geared for war', or, to use Wilson's terminology, 'politically militarized'. There is also reason to believe that the Shah's rule was dependent upon military success (Whitby 1994), and that an heroic ideal and its concomitant warrior ethos were stressed by the Sasanid court, and internalized by the provincial elite (Walker 2006, 122 ff.). Given that there is general agreement that the elites of these neighbouring kingdoms and tribes were militarized, the same might be true for East Rome.

War, Imperial Propaganda, and the Elite Imagination

Having defined the key terms in this discussion, and looked briefly at some suggestive comparative evidence, what remains are the East Roman ruling elite themselves. I begin with signs of Wilson's 'cultural militarization' amongst the upper echelon of this group: the emperors.

Military success always played an important part in the success and imagery of an emperor (Campbell 1984, Whitby 2004), and after a brief hiatus in the 5th century AD, there is a return to this militaristic ideology, with McCormick (1986, 67) noting that 'Justinian's reign witnessed a powerful reassertion of the image of the victor emperor'. One particular medium through which military success could be extolled was literature (McCormick 1986, 35-130; Lee 2007, 37-50). Priscian (*Panegyric* 63-65) praises Anastasius' success against the Isaurians. In the introduction to his collection of epigrams Agathias (*Greek Anthology* 4.3) celebrates Justinian's victories. Procopius' *Buildings* spends a great deal of time on fortifications which he claims were built by Justinian. John the Lydian (John Lydus, *De magistratibus* 3.28; McCormick 1986, 64, n.100; Greatrex 1998, 61) was asked by Justinian to compose a history of his Persian wars, although this is no longer extant. An emperor need not have had considerable military success to merit such attention, as Corippus (*In laudem Iustini minoris* 2.105-127) praises Justin II for his success against unnamed foes. Occasionally an emperor would praise his own exploits, for in Justinian's *Digest* we find a huge list of victory titles: 'Emperor Caesar Flavius Justinian Alamannicus, Gothicus, Francicus, Germanicus, Anticus, Alanicus, Vandalicus, Africanus, Pious, Forutnate Glorious Victor and Triumphator Ever Augustus' (*Institutiones* preface 1).

Besides the literature, military imagery also found its way into public art. An enormous bronze equestrian statue in Constantinople, which is no longer extant is ascribed to Justinian by Procopius in his *Buildings* (1.2.5-12):

> On this statue sits an equally massive bronze emperor wearing a Homeric breastplate as well as a costume called Achilles. In his left hand is a globe with a cross on it, which, together, symbolize his victory in war and mastery of the world, while his right hand is held out to the east, so bidding the peoples there to advance no further.

Procopius also includes in his *Buildings* a description of a mosaic – which again, is no longer extant – on the roof of the entrance vestibule to Justinian's palace, the Chalke:

> This mosaic is filled with scenes of war ('*polemos*') and battle ('*mache*'), the capture of cities in Italy and Africa, the general Belisarius giving spoils to the emperor, who stands in the middle with the empress Theodora, with the king of the Vandals and the king of the Goths being received by the imperial couple, while surrounding the lot are the senate. (Procopius, *Buildings* 1.10.16).

Other means were also used to convey these images of victory to the public. Justinian often held his celebrations in the Hippodrome, including parades (Marcellinus Comes, 103.7-10; Croke 1980, 2005, 78). Sometimes there were grand ceremonies to see off an expeditionary army, such as at the launching of the Vandal expedition. Procopius (*Wars* 3.11) describes this, providing, in Homeric fashion, a catalogue of ships. The expeditionary forces anchor off the point by the royal palace, where the emperor sees them off, but not before the mission is blessed by the chief priest of the city Epiphanius (Procopius, *Wars* 3.12.1-2).

Personal items also conveyed similar imagery. The Barberini ivory or diptych (Kitzinger 1977, 96-97, plate 176) depicts an emperor (presumably Justinian) in the centre panel mounted on a horse – we are reminded of Justinian's equestrian statue – accompanied by a winged victory, and with a high-ranking military officer on the left panel offering the emperor a statue of another winged victory. Corippus (*In laudem Iustini minoris* 3.120-125) describes golden tableware which Justinian is said to have had inscribed with scenes of his Vandal triumph. Corippus (*In laudem Iustini minoris* 1.275-287) also claims that Justinian had fine clothing embroidered with images of victory, including a depiction of himself trampling upon the Vandal king (McCormick 1986, 68; Lee 2007, 82).

Emperors and imperial imagery aside, war and military affairs seem to have inspired a number of writers, with a considerable number of secular works having a discernable military character produced in the 6th century AD. The Greek classicizing histories of Procopius and Agathias are dominated by warfare to a much greater extent than the works of earlier classicizing historians such as Herodian, though we are regrettably ill-informed about the works of authors such as Dexippus, Eunapius, Olympiodorus, and Priscus. The case of Procopius is perhaps unsurprising, as he was an eye-witness and participant for much of the action he describes, and parts of his *History of the Wars of Justinian* include some very heroic descriptions (Whately 2009). Agathias is not known to have had any experience of combat, yet decided to focus his work on warfare. Although the remaining 6th century AD history, that of Menander Protector, only survives in fragments, Brodka (2007) has postulated that war greatly affected the historian, shaping the world-view evident in the text, with fear, threats, and the fall of empire being important themes of the work. Indeed, Brodka's (2007, 100) comments are very much in keeping with what has been said earlier about the military character of the period and East Rome's neighbours. Classicizing historiography aside, a number of military treatises were also written in the 6th century AD, to some degree rehabilitating a genre which had not been well-served in the preceding century or more. This includes the *Strategikon* of Maurice, the *Epitedeuma*, and the *Tacticon* of Urbicius, and possibly the *Peri Strategias*, the *Rhetorica Militaris*, and the *Naumachia* of Syrianus, though these might date to the 10th century AD. Moving on the anonymous author of the *Dialogus de scientia politica* includes a debate about cavalry and infantry, while Evagrius' *Ecclesiastical History*, and John Troglyta's *Iohannidis*, an epic poem written in Latin, also discuss military matters. The *Chronicle* of Marcellinus Comes is filled with military concerns, as is Jordanes' *Getica*. The same can be said about John the Lydian's *On Magistrates*, in which considerable space is devoted to military issues. All of these works found an audience, and their very diversity points towards the existence of a sizeable portion of the elite with an interest in the military.

Literature aside, military matters found their way into the consciousness of some individuals in other ways. Belisarius, for example, was much loved by the people of Constantinople due to his military record, and this is particularly evident in his return to the capital following his first expedition to Italy. Procopius (*Wars* 7.1.4) notes that 'everyone was talking about Belisarius, having been attributed two victories, such as no man had happened to manage before', and later that 'it was a pleasure to the Byzantines to see Belisarius' (Procopius, *Wars* 7.1.5). Although much of the attention alluded to here hails from the peasants, their attention is suggestive. Procopius also includes some interesting comments about contemporary reactions towards the various military appointments of the general Bessas, who struggles when placed in charge of Rome, returns to Constantinople, and is dispatched as general of the East by Justinian. Procopius notes (*Wars* 8.12.33): 'nearly everyone bitterly criticized this decision', and a line later (*Wars* 8.12.34) 'but though nearly everyone thought this'. Unfortunately it is not known to whom this 'nearly everyone' refers. It could conceivably be the royal we, that is, the author Procopius alone, although it could also be Procopius' peers. Regardless, these two episodes point towards the attention which commanders apparently attracted amongst a sizeable portion of the public.

There is other evidence that military issues were the focus of many 6th century AD discussions among the elite. Some scholars (Greatrex, Elton, and Burgess 2005, 70-71; Kaldellis 2004, 190-204) have drawn attention to what is perceived as a contemporary tactical debate as to whether cavalry or infantry were to be preferred in battle. Urbicius, in his *Epitedeuma*, seems to feel that the infantry still had an important role to play, while Maurice, by originally focusing on cavalry in the *Strategikon*, seems to suggest that infantry was not in need of improvement. The anonymous author of the *Dialogus de scientia politica* (4.28-53) stresses the importance of cavalry. The same may be true of John the Lydian (*De magistratibus* 1.14), who assigns an important role to the *magister equitum*. It is certainly true of Procopius (*Wars* 1.1.6-16), who in the preface to his *Wars* contrasts contemporary horse-archers with their Homeric forebears, so arguing that the newer models are far superior, in essence the ideal warriors. Ironically, Belisarius, the central character of Procopius' *Wars*, is somewhat unsure of the value of cavalry (Procopius, *Wars* 5.28.21-29). Such debates are not found in earlier periods of Roman history, and notably not in those periods which are said to have an undeniably militaristic character.

Military Careers

The previous discussion, though little more than an outline, does suggest there was a considerable degree of cultural militarization in 6th century AD East Rome. What has not yet been explored is societal militarization, which was a prominent feature of Republican Rome. During the Middle Republic, military success was the true measure of political success (Polybius, 6.19.4; Harris 1979, 11; Serrati 2007, 486). Aspiring young elites would work their way up through the *cursus honorum*, from quaestor to consul, with success on the field of battle the indicator for most Romans of worthiness to hold higher public office. A good indicator of societal militarization in 6th century AD East Rome, then, would be evidence of individuals who had served in the armed forces before taking up political or civilian office.

The highest ranks in the military were the *magistri militum*, who commanded the mobile *comitatenses* field armies, which were themselves divided into two categories, the central armies and the regional armies. There were several of these: one for the northeast, particularly Armenia; one for the southeast; one for Africa; one for Italy; one for Thrace and Illyricum; and two for Constantinople and its vicinity. The literary sources are summarily vague on the command structure, with many historians simply referring to commanders as *strategoi* and *archontes* (Jones 1964, 654-656; Bréhier 1970, 274-275; Ravegnani 1988, 73-75; Whitby 2000a, 289; and Ravegnani 2004, 45-47). Below them were the *duces*, who were in charge of the frontier armies, the *limitanei*, and the *comites rei militaris*, who often held special commands (Jones 1964, 656; Bréhier 1970, 275; Ravegnani 1988, 74-75; and Whitby 2000a, 288-290). Different grades within society reflected one's position. These were, from highest to lowest, *illustris*, *spectabilis*, and *clarissimi* or *perfectissimi*. With respect to the soldiery, a *magister militum* and *exarch* (formerly the office *magister militum*) were of the *illustris* grade, while *duces* were *spectabilis* (Ravegnani 1988, 76-82).

The best repository of information regarding military careers in the 6th century AD is the *Prosopography of the Later Roman Empire (PLRE)*. Nothing else provides such a detailed body of data. Based on my tabulations of individuals dating to the 6th century AD in the third volume of the *Prosopgraphy of the Later Roman Empire*, there are 953 such people, including three emperors. It is likely that some of the 953 are duplicates, particularly amongst the Ioannes. Roughly one third (301 individuals) of the total are elite soldiers, that is, highly ranked officers, though I have included those with honorific titles. Of this group of officers, 53 are of the grade *spectabilis*, while the rest (248 individuals) are *illustris*, accounting for more than eight-tenths of the total. As regards the civilians (653 individuals), 81 are of the *spectabilis* grade, with the rest (572 individuals) being *illustris*, a little less than nine-tenths of the total. The low number of those of *spectabilis* grade is probably because the title gradually lost its lustre over the course of late antiquity (Jones 1964, 529). Overall, more than two-fifths (248/572) of those of the grade *illustris* are soldiers. Consequently, although hardly conclusive, it seems that a significant portion of the highest ranking group of elites (though, inevitably, not as high as the *PLRE* suggests) served in the armed forces.

To complement this picture, we need to know more about individual career patterns, and particularly whether members of the ruling elite who had held supreme military commands later went on to hold important civilian offices. Family history would also be instructive, particularly if there are cases of individuals who attained important political office after their fathers, uncles, brothers, and so forth had themselves been high-ranking officers. By their very nature, those who held the highest military commands would be considered ruling elites because of their holding of the rank of *illustris*. The question is whether this led to other types of success, as was the case during the Republic, or whether these elites functioned on a level parallel to those holding political offices. For the most part our evidence is insufficient, though there are exceptions. For example, we do have evidence of the reverse: those with civilian backgrounds, often with no apparent military experience, given a major military command, such as Areobindus (2). Members of the imperial family, such as Justin I's nephew Germanus (*PLRE* III, 527), would regularly be given such offices (Ravegnani 1988, 87). The careers of Areobindus and his kin, and that of Buzes, whose exploits are described by Procopius, are worth discussing in some detail, due to the evidence they provide of the place of military careers in late antiquity.

Areobindus was a *magister militum* in Africa in AD 545 (Procopius, *Wars* 4.24.1), and his nomenclature suggests that he was probably related to Flavius Areobindus Dagalaiphus Areobindus (1, *PLRE* II, 143 ff.) who was consul (honorary) in AD 506, the Flavius Ariobindus who was consul in AD 434 (2, *PLRE* II, 145 ff.), and, necessarily, the Flavius Dagalaiphus (2, *PLRE* II, 340) who was consul in AD 461. Fl. Areobindus Dagalaiphus Areobindus, who, quite possibly (based on his name and his chronology), had been Areobindus' father, had been a *magister militum per Orientem* in AD 503-505 during Rome's war with Persia (Pseudo-Joshua Stylites, 281). Before holding that office he had been count of the stables, and his ordinary consulship came not long after his military command (*PLRE* II, 143). The former office was a civilian one, despite its relevance to the military (Jones 1964, 625-626), and Fl. Areobindus Dagalaiphus Areobindus' career was not unusual, as Flavius Iordanes (3, *PLRE* II, 620) had been *comes stabuli* before later taking up the offices of *magister militum per Orientem* (AD 466-469) and *consul* (AD 470).

Fl. Areobindus Dagalaiphus Areobindus was the son of Fl. Dagalaiphus (2, *PLRE* II, 340), who himself had been consul and was a patrician. Our information about him is limited, and at present there is no evidence that he had served in the military, although it is possible that he did. Fl. Dagalaiphus' father Fl. Ariobindus (sic) had been a *magister militum* in the East and, concurrently, a consul, and before that a *comes foederatorum* (2, *PLRE* II, 145). His parentage is unknown. There was, however, a Dagalaifus (sic) who had fought against the Alamanni in Gaul in the middle of the 4th century AD (Ammianus Marcellinus, 27.2.1 ff.). Although there is no proof that the former is descended from the latter, there is a possibility that this family had a tradition of military service.

Ignoring the 4th century AD Dagalaifus, several members of this family served as high ranking military commanders and were often also consuls, with the pattern only broken by Areobindus 2, the master of soldiers in Africa. As noted above, this man does not seem to have had any military experience prior to being sent to Africa, although the evidence is far from conclusive. If we assume that he is the son of the Fl. Areobindus Dagalaiphus Areobindus who had been a commander during the war with Persia in the early 6th century AD, then we have here a family whose long tradition of military service had resulted in the attainment of patrician status. This is less likely, of course, if his father is the Dagalaiphus who was a *comes Aegypti* in AD 509 (1, *PLRE* II, 340), although not entirely so, especially if this Dagalaiphus is the father of Areobindus 4 (*PLRE* III, 110). Even if Areobindus 2, Areobindus 4, and, for that matter, Areobindus 3 are all the same person (which is a possibility, however remote), then if they are also sons and grandsons of the aforementioned masters of soldiers, the point holds: distinguished military service can bring high civilian office, though not necessarily in one generation.

The career trajectory for the various members of this family may have been exceptional. However, there is at least one other family which had a distinguished record of military service, and which also had members who attained a high civilian office, although they are fewer in number. Buzes was *magister militum per Orientem*, and later a *magister magistrum* in Lazica, having been a *dux* with his brother Cutzes in Lebanon (Procopius, *Wars* 1.13.5; *PLRE* III, 255). Like Areobindus 2, the precise identity of Buzes' father is unknown, though Flavius Vitalianus has been suggested by Martindale (2, *PLRE* III, 255). The evidence for this suggestion comes from Malalas (Malalas 441), and is not corroborated by Procopius or Zachariah, which is a reason for concern (Cutzes, *PLRE* III, 366), although, to my mind, unnecessarily so. Vitalian had a distinguished military career, serving as a *magister militum per Thracias* and a *magister militum praesentalis* (2, *PLRE* II, 1171). In AD 518 he was acclaimed as a *patricius* (*Acta Conciliorum Oecumenicorum* III, 85.26, 86.21; *PLRE* II, 1175), and two years later, though briefly, held the consulship (Jordanes, *Romana* 361; Marcellinus Comes, 520; Malalas, 18.26; Evagrius, 4.3). Vitalian's father Patriciolus had been a military man himself, and served in the same war as Fl. Areobindus Dagalaiphus Areobindus, though possibly as a *comes foederatorum* (*PLRE* II, 837). Unfortunately, we know nothing about Patriocolus' parentage.

Buzes had at least two brothers. One, Benilus, was a commander in Lazica (*PLRE* III, 224-225), while the other, Cuzes, was *dux* of Phoenicia (*PLRE* III, 366). The three brothers, Buzes, Benilus, and Cuzes in turn had a nephew named Domentiolus. Domentiolus, like his uncles, was a military man though he seems never to have achieved any senior military commands. He did, however, lead Roman forces during the war with Sasanid Persia in the AD 540s, and later commanded troops in Sicily against Totila and the Goths (*PLRE* III, 413). Domentiolus had a son named Ioannes (81) whose career seems largely to have been that of a civilian. In the middle of the 6th century AD he was a financial official in Constantinople, and later a *patricius*, who served as an ambassador to Persia (*PLRE* III, 672-673).

There are two other possible members of this family, Bonus (2, *PLRE* III, 241) and Ioannes (46, *PLRE* III, 652). Bonus was either the cousin or nephew of Ioannes. Ioannes, in turn, was the nephew of a Vitalian, seemingly the same Vitalian who was the probable father of Buzes and his siblings (Procopius, *Wars* 6.5.1, 6.7.25, 6.28.33, 7.3.2, 7.5.4, 7.34.41, 7.39.10, 7.40.10, 8.23.4, 8.34.22). Again, Martindale (*PLRE* III, 241, 652) is more cautious than necessary, and there is no reason to question the familial relationship of these two men with Buzes and his kin. Bonus seems to have had a minor military command, while John was the *magister militum per Illyricum*. There is every reason to believe that the latter's success was largely due to his marriage to Justina, the daughter of Justinian's cousin Germanus (*PLRE* III, 652). It seems probable that he later became *patricius*, and possibly also honorary consul (Ioannes 71, 72, *PLRE* III, 669, 670). To summarize, although it took several generations, the family of Buzes, like that of Areobindus, had members who attained important civilian careers, seemingly on the back of their seniors' distinguished military careers and the wealth that may have stemmed from them.

Conclusion

The stated aim of this paper was to address the question of whether there was a militarization of the elite or the rise of a military culture in the 6th century AD. I have equated the 'militarization of the elite' with Wilson's societal and cultural militarization, explained my preference for the term 'elite' over 'aristocracy' and stated that the term 'Roman' was being used in the broadest sense of the word, so obviating problems concerning other ethnic groups. I have highlighted some evidence suggesting that the ruling elites in the 6th century AD were concerned with warfare, which is to say that they were culturally militarized, something which was not confined to those with a military background. Consequently, the first half of my central question seems to have been answered in the affirmative.

I then turned to military careers, the hallmark of societal militarization in the mid-Republic, and highlighted the careers of two relatively prominent families, that of Areobindus and Buzes. Although many important details about these two families are unknown, their respective military backgrounds do seem to have led, after some time, to high ranking civilian offices, and in particular, the consulship. Here the evidence was far from conclusive, and seemed to support the notion of the rise of a military culture. For much of the period under consideration, members of the respective families did not attain political office. Indeed, the experiences of the family of Buzes led Ravegnani (1988, 90) to posit the existence of a pseudo-military aristocracy in the age of Justinian. In an important study on government in the 5th and 6th centuries AD, Barnish, Lee, and Whitby (2000) suggested that the state was effectively run by a polycracy, that is, by elites of all types, from civilian elites, to military and clerical elites, working in parallel, and sometimes in opposition. For those scholars, the military elites began to take a more prominent role during the reign of Maurice, a period of profound change (Barnish, Lee, and Whitby 2000, 170). Returning to the two families, their career paths were very much military until the middle of the 6th century AD, when civilian offices became a real possibility. None, however, took the three main palatine offices, arguably the highest of civilian offices (Barnish, Lee, and Whitby 2000, 171), the *comes sacrarum largitionum*, the *comes rerum privatarum*, and the *magister officiorum*, so suggesting that some sort of division still existed, though a select few did attain the consulship, the ultimate reward for distinguished service in the military and palatine offices (Jones 1964, 532-533). The question of societal militarization must, then, necessarily remain unanswered, though there are hints that this process may have begun during the reign of Justinian. A much more detailed study of civilian and military careers is needed before any firm conclusions can be drawn. More work is also needed on the political power of commanders in the 6th century AD East Roman Empire.

Finally, this paper has not discussed Wilson's concepts of political militarization or economic militarization. Suffice it to say that the titles assumed by Justinian and his reconquest of the west are all suggestive, and although diplomacy was often an option in foreign affairs, every 6th century AD emperor was willing to go to war, albeit not necessarily in person. Indeed, both Anastasius and Justinian engaged in a significant amount of fortification work in the Balkans and in the eastern frontier regions. As regards economic militarization, the willingness of emperors to engage in warfare and spend significant amounts constructing the physical and social structures of war suggests that it was most certainly present. Nevertheless, these issues, too, require further study.

Bibliography

Amory, P. 1997. *People and Identity in Ostrogothic Italy*. Cambridge, Cambridge University Press.

Barnish, S., Lee, D. and Whitby, M. 2000. Government and administration. In A. Cameron, B. Ward-Perkins and M. Whitby (eds.), *The Cambridge Ancient History Volume XIV*, 164-206. Cambridge, Cambridge University Press.

Bréhier, L. 1970. *Les Institutions de L'Empire Byzantin*. Paris, Albin Michel.

Börm, H. 2010. Herrscher und Eliten in der Spätantike. In H. Börm and J. Wiesehöfer (eds.), *Commutatio et Contentio. Studies in the Late Roman, Sasanian, and Early Islamic Near East*, 159-198. Dusseldorf, Wellem Verlag.

Brodka, D. 2007. Zum Geschichtsverständnis des Menander Protektor. *Electrum* 17, 95-103.

Campbell, B. 1984. *The Emperor and the Roman Army, 31BC-AD 235*. Oxford, Oxford University Press.

Canepa, M. 2009. *The Two Eyes of the Earth. Art and Ritual Kinship Between Rome and Sasanian Iran*. Berkeley, University of California Press.

Cornell, T. 1993. The End of Roman Imperial Expansion. In J. Rich and G. Shipley (eds.), *War and Society in the Roman World*, 38-68. London, Routledge.

Croke, B. 1980. Justinian's Bulgar Victory Celebration. *Byzantinoslavica* 41, 188-195.

Croke, B. 2001. *Count Marcellinus*. Oxford, Oxford University Press.

Croke, B. 2005. Justinian's Constantinople. In M. Maas (ed.), *The Age of Justinian*, 60-86. Cambridge, Cambridge University Press.

Eich, P. 2007. Militarisierungs- und Demilitarisierungstendenzen im dritten Jahrhundert n. Chr. In L. de Blois, and E. Lo Cascio (eds.), *The Impact of the Roman Army (200 BC - AD 476)*, 511-515. Leiden, Brill.

Greatrex, G. 1998. *Rome and Persia at War, 502-532*. Leeds, ARCA.

Greatrex, G., Elton, H. and Burgess, R. 2005. Urbicius, *Epitedeuma*: An edition, translation and commentary. *Byzantinische Zeitschrift* 98, 35-74.

Goffart, W. 2006. *Barbarian Tides*. Philadelphia, Pennsylvania University Press.

Grosse, R. 1920. *Römische Militärgeschichte von Gallienus bis zum Beginn der byzantinischen Themenverfassung*. Berlin, Wiedmann.

Haldon, J. 2004. Introduction: Elites Old and New in the Byzantine and Early Islamic Near East. In J. Haldon and L. Conrad (eds.), *The Byzantine and Early Islamic Near East VI: Elites Old and New in the Byzantine and Early Islamic Near East*, 1-12. Princeton, Darwin Press.

Haldon, J. and Conrad, L. (eds.) 2004. *The Byzantine and Early Islamic Near East VI: Elites Old and New in the Byzantine and Early Islamic Near East*. Princeton, Darwin Press.

Halsall, G. 2003. *Warfare and Society in the Barbarian West, 450-900*. London, Routledge.

Halsall, G. 2007. *Barbarian Migrations and the Roman West 376-568*. Cambridge, Cambridge University Press.

Harris, W. 1979. *War and Imperialism in Republican Rome 327-70 BC*. Oxford, Oxford University Press.

Heather, P. 1996. *The Goths*. Oxford, Blackwell.

Howard-Johnston, J. 2008. State and Society in Late Antique Iran. In V. S. Curtis and S. Stewart (eds.), *The Idea of Iran*, III *The Sasanian Era*, 118-131. London, I. B. Tauris.

Howard-Johnston, J. 2010. The Sasanians' Strategic Dilemma. In H. Börm and J. Wiesehöfer (eds.), *Commutatio et Contentio. Studies in the Late Roman, Sasanian, and Early Islamic Near East*, 37-70. Dusseldorf, Wellem Verlag.

Hunt, P. 2010. Athenian Militarism and the Recourse to War. In D. M. Pritchard (ed.), *War, Democracy and Culture in Classical Athens*, 225-242. Cambridge, Cambridge University Press.

Jones, A. H. M. 1964. *The Later Roman Empire 284-602*. Oxford, Blackwell.

Kaldellis, A. 2004. Classicism, Barbarism, and Warfare: Prokopios and the Conservative Reaction to Later Roman Military Policy. *American Journal of Ancient History*, n.s. 3, 189-218.

Kaldellis, A. 2007. *Hellenism in Byzantium*. Cambridge, Cambridge University Press.

Kitzinger, E. 1977. *Byzantine Art in the Making*. London, Faber.

Kouroumali, M. 2005. *Procopius and the Gothic War*. Unpublished DPhil Dissertation, University of Oxford.

Kuefler, M. 2001. *The Manly Eunuch. Masculinity, Gender Ambiguity, and Christian Ideology in Late Antiquity*. Chicago, University of Chicago Press.

Lee, D. 2007. *War in Late Antiquity*. Oxford, Blackwell.

Lendon, J. E. 1997. *Empire of Honour*. Oxford, Oxford University Press.

MacMullen, R. 1963. *Soldier and Civilian in the Later Roman Empire*. Cambridge MA, Harvard University Press.

Martindale, J. R. 1980. *The Prosopography of the Later Roman Empire Volume II, AD 395-527*. Cambridge, Cambridge University Press.

Martindale, J. R. 1992. *The Prosopography of the Later Roman Empire Volume III, AD 527-641* (PLRE III). Cambridge, Cambridge University Press.

Mattern, S. 1999. *Rome and the Enemy*. Berkeley, University of California Press.

McCormick, M. 1986. *Eternal Victory*. Cambridge, Cambridge University Press.

Mitchell, S. 2007. *The History of the later Roman Empire*. Oxford, Blackwell.

Müller, A. 1912. Das Heer Justinians nach Procop und Agathias. *Philologus* 71, 101-138.

Pollard, N. 1996. The Roman army as 'Total Institution' in the Near East? Dura-Europos as a case study. In D. L. Kennedy (ed.), *The Roman Army in the East*, 211-227. Ann Arbor MI, Journal of Roman Archaeology.

Potter, D. 2004. *The Roman Empire at Bay AD 180-395*. London, Routledge.

Rance, P. 2005. Narses and the Battle of Taginae (Busta Gallorum): Procopius and Sixth-Century Warfare. *Historia* 54, 424-472.

Ravegnani, G. 1988. *Soldati di Bisanzio in età giustinianea*. Rome, Jouvence.

Ravegnani, G. 2004. *I Bizantini e la Guerra*. Rome, Jouvence.

Salzman, M. R. 2002. *The Making of a Christian Aristocracy*. Cambridge MA, Harvard University Press.

Serrati, J. 2007. Warfare and the State in Hellenistic Greece and the Roman Republic. In P. Sabin, H. Van Wees and M. Whitby (eds.), *Cambridge History of Greek and Roman Warfare Volume I*, 461-497. Cambridge: Cambridge University Press.

Shaw, B. 1999. War and Violence. In G. Bowersock, P. Brown and O. Grabar (eds.), *Late Antiquity: a Guide to the Postclassical World*, 130-169. Cambridge MA, Harvard University Press.

Teall, J. 1965. The Barbarians in Justinian's Armies. *Speculum* 40, 294-322.

Van Dam, R. 2005. Merovingian Gaul and the Frankish Conquests. In P. Fouracre (ed.), *The New Cambridge Medieval History Volume I, c. 500-700*, 193-231. Cambridge, Cambridge University Press.

Walker, J. T. 2006. *The Legend of Mar Qardagh. Narrative and Christian Heroism in Late Antique Iraq.* Berkeley, University of California Press.

Whately, C. 2009. *Descriptions of Battle in the* Wars *of Procopius*. Unpublished PhD Dissertation, University of Warwick.

Whitby, M. 1994. The Persian King at War. In E. Dabrowa (ed.), *The Roman and Byzantine Army in the East*, 227-263. Krakokw, Drukarnia Uniwersytetu Jagiellonskiego.

Whitby, M. 1995. Recruitment in Roman armies from Justinian to Heraclius, ca. 565-615. In A. Cameron (ed.), *The Byzantine and Early Islamic Near East III: States, Resources and Armies*, 61-124. Princeton, Darwin Press.

Whitby, M. 2000a. The Army, c. 420-602. In A. Cameron, B. Ward-Perkins and M. Whitby (eds.), *The Cambridge Ancient History Volume XIV*, 288-314. Cambridge, Cambridge University Press.

Whitby, M. 2000b. Armies and Society in the Later Roman World. In A. Cameron, B. Ward-Perkins and M. Whitby (eds.), *The Cambridge Ancient History Volume XIV*, 469-495. Cambridge, Cambridge University Press.

Whitby, M. 2004. Emperors and Armies, 235-395. In S. Swain and M. Edwards (eds.), *Approaching Late Antiquity: The Transformation from Early to Late Empire*, 156-186. Oxford, Oxford University Press.

Wickham, C. 2005. *Framing the Early Middle Ages.* Oxford, Oxford University Press.

Wilson, P. 2008. Defining Military Culture. *The Journal of Military History* 72, 11-41.

Malice in Wonderland:
The Role of Warfare in 'Minoan' Society

Barry P. C. Molloy

Abstract

This paper offers diachronic observations on the character and role of warfare and violence in the societies of Bronze Age Crete in light of recent developments in the study of warfare in prehistory. The main emphasis is on the functional properties of weapons and how these related to combat practices from the perspective of individual combatants rather than the macro-perspective of armies and tactics. The bodies of material culture examined are determined temporally as c.3000-1100 BC and spatially as the island of Crete, with reference to the Argolid on mainland Greece where required. Cretan societies are revealed as possessing complex military institutions supported ideologically and technologically by elite frameworks, and violence and warfare are thus revealed as systemic aspects of these societies. The polarity between 'peaceful Minoans' and 'warlike Mycenaeans' that often underlies social models implicitly, if not explicitly, is questioned on the grounds that perpetual two-way transfers of military technology existed between the two regions from the origins of purpose-made combat weaponry to the end of the Bronze Age. Furthermore, it is argued that the political, economic and technological capacities of Cretan societies between 3000 and 1450 BC were a more favourable environment than mainland Greece for the origins of the military equipment, practices and ideologies found widely throughout the southern Balkan peninsula and Aegean islands.

Introduction

Whether Minos ruled the waves or Agamemnon led a confederation against Troy, the archaeological visibility of warlike material culture and imagery in the Aegean surpasses most areas of Europe in quantity and detail. The popular myth of 'peace-loving' Minoans may be occasionally espoused by enthusiastic tour-guides at archaeological sites, but if it ever truly existed in academic circles, it is not a dominant model. Crete, however, is perhaps unique in European prehistory in that there has to date been no comprehensive treatment of the role of war and violence in society. As this island saw the first manifestation of urbanism in Europe and has long been used as a testing ground for models of the evolution of social complexity, the key role warfare is known to have played in other societies needs to be better reflected in our conception of Cretan societies through time. This paper does not claim to be a complete and thorough assessment of the evidence, as it would have to be dramatically longer to undertake this, but it will rather seek to re-contextualise data that may be used to assess the relationship between war and society.

A further ambition of the paper will be to reconsider the relationship between Crete and the Mainland (MacGillivray 2000) in military terms during the Bronze Age, and in so doing, it will challenge what may be better considered a host of factoids rather than paradigms in Aegean research. Though rarely explicit, regional polarity in our appreciation of military systems can be seen to permeate most macro-scale analyses of the complex networks of power and identity in Aegean prehistoric societies (see various contributions in Shelmerdine 2008). It is hoped that this exploration of war and society in prehistoric Crete may dispel any lingering myths of peaceful Minoans (Bintliff 1984) while offering a more coherent framework for considering the shared martial heritage of Aegean Bronze Age societies. Taking a diachronic view of Crete in the Bronze Age, it will be argued that warfare continually played a systemic role in socio-political dynamics.

Daggers, Swords and Legitimacy – EM II to MM II

The role of defensive architecture in this time span and during the Neopalatial period has been the subject of research for some time (Alexiou 1979; Starr 1984; Tzedakis et al. 1990; Driessen and Macdonald 1997; Chryssoulaki 1999; Alusik 2007). It will not be considered in detail in this paper because as yet we have no evidence to suggest that they dominated the urban environment or had significant impact on patterns of warfare. The multiple non-martial readings of defensive architecture, and the bias in its documentation towards lower order settlements (Alusik 2007) diminishes its validity for the present discussion.

The actions of combat rarely provide long-term physical evidence of warfare on the field of battle leaving us few resources to analyse the nature of such events. The prehistory of warfare is thus impoverished by the invisibility of its direct manifestation, its materiality is often best expressed through the tools used in its undertaking and we are lucky that cupriferous weapons often survive in remarkably good condition. In the early stages of the Bronze Age it can be difficult to differentiate the weapons of war from objects suited to other contexts, hunting in particular. As the use of daggers in violent activities is however not in contention, allowing for their use in combat ties them in with expressions and perceptions of male identity which includes characteristic elements of warriorhood.

We can note that the possession of a dagger in burials affords a martial flavour to the construction of identity in death in the EM and early MM periods, an element echoed some time later on the male figurines from some Peak Sanctuaries who carry such weapons (Peatfield 1990). The conspicuous depiction of a weapon on a male figurine where so few other non-anatomical details are emphasised is an important indicator as to the perceived importance of martial symbolism for a Cretan audience. Peatfield (1999) has characterised these daggers as being well suited to combat contexts, and while they served other potential functions, the great length of some of the Aghia Photia pieces in particular emphasises the martial capacities of these early weapons (see experimental work in Molloy 2006, *contra* Branigan 1999). In this Prepalatial and early Protopalatial world the possession of a dagger and/or spear and the ability to use them effectively appears to have been a widespread phenomenon. The function of spears as markers of identity is hinted at in some of the rare Protopalatial figural scenes, but a most remarkable sealstone of EM III or MM I date illustrates an unambiguous scene of a dagger being used in interpersonal combat (Hatzi-Vallianou 1979; Wedde 2000; Papadopoulos 2006).

One can see that many adult males had access to the tools needed to defend their communities and properties, or conversely to aggressively subjugate or loot other groups on a relatively small regional scale (though not exclusively small). Branigan (1999, 88) has remarked that over 80% of bronze objects from Early Bronze Age Crete are weapons, with some 73% of all Early and Middle Bronze Age Aegean daggers coming from the island (Peatfield 1999, 67). By MM II important developments in martial material culture occur at the same time as increasing centralisation of power networks in particular regions of Crete takes place, often around the growing urban centres (Watrous 2001; Schoep 2006; Haggis 2007; Knappet 2008). Urban developments were but one element to increasing territorialism and while there is a lack of hard evidence for or against the role of coercive force in this process, it is hard to conceive of a model whereby the protection of newly acquired wealth is by divine mandate alone. Human nature has never been such that independent neighbouring groups can control resources and enrich themselves materially with mutual respect and understanding (Keeley 1996). The weapons finds from Mallia (Kilian-Dirlmeier 1993) indicate another side to this story, one of violence which is far more typical of all other known historical cultures (Keegan 1994).

Two MM II swords of Type A were found at Mallia, and represent our earliest examples of true swords in the Aegean (Kilian-Dirlmeier 1993). Technologically they represent a significant divergence from the so-called 'Crystal Pommel' sword found close by (Sandars 1961, 19; Kilian-Dirlmeier 1993, 15), as these were undoubtedly functional weapons suited to use in interpersonal combat (Peatfield 1999, 2007; Molloy 2006; 2008; in preparation). It is interesting that the other finds from this area also have direct martial symbolism, from the elaborately decorated dagger to the stone macehead in the form of a leaping leopard transforming into an axehead at the tail end. Of more importance still is the repousse decorated golden hilt furniture found with the swords. This is normally interpreted as a leaping acrobat (Chapouthier 1938), but on closer inspection it may well be seen as a captive bound at the wrists and ankles curled around the central perforation, its non-Cretan short curly hair identifying it as a subjugated foreigner. Given the comparative dearth of figural art in this period it is interesting to note the aggressive undertones of this image.

The Type A swords that we find at Mallia are not longer versions of daggers (Molloy 2010). They are the first true swords and represent therefore the first definite evidence for the art of swordsmanship in the Aegean. This development is considerably more profound than it appears at face value as the evolution of swords and swordsmanship represents the development of weapons specifically designed to be used in the context of interpersonal combat. There is no ambiguity in this as they are ill-suited to any other functions, even hunting, as determined through innumerable parallels (Oakeshott 1999; Clements 2007). Swordsmanship required the development of significant new skill sets, as the manner in which they can be effectively used requires martial expertise that can only be accrued through concerted training (Molloy 2006; 2008).

The technology to produce daggers is comparatively simple due to their short length, but the casting of these much longer and thinner objects requires developments in both mould making and pyrotechnology in order to feed the bronze the full length of the mould (Burridge personal communication). Thus these first swords represent the evolution of complex bronze working techniques (Weiner 1989) and the evolution of a new combat system which facilitates the very specific requirements needed for the effective use of these swords. The time, skill and material investment in this process could best be supported by the growing urban centres as evidenced by later known weapon foundries at Knossos (Driessen and Macdonald 1984; Kilian-Dirlmeier 1993). Access to the tools of combat and the training to use them skilfully were thus increasingly circumscribed, controlling access to important developments in warfare. This move was the beginning of a process which supplanted older martial traditions by MM III, centralising the control of legitimate violence to the elite groups.

While Acts of God may have brought about the systematic widespread destructions in the closing years of the 18th century BC, increasing social tensions (Schoep 2006) manifesting themselves through open warfare and novel martial developments are also likely sources for some components of this protracted period of destruction. Nowicki has argued that the so-called 'refuge sites' well known in the LM IIIC period also saw quite extensive occupation in late MM II, particularly close to natural territorial boundaries (Nowicki 2008, 77-80). That they were rapidly depopulated again in MM III indicates that the upheavals around 1700 BC had passed and that people no longer needed defensible settlements, or perhaps that central authorities of the emerging Neopalatial system imposed more easily manageable settlement dynamics. The decades late in the ceramic phase MM II and early in MM III were times of intense disruption in the centre and east of the island at least. If warfare was an aspect to the destruction of older power systems, there can be little doubt that the control of violence by the elite would be paramount in developing new and more stable centres of power and wealth in the ensuing centuries.

The Plot Thickens – MM III to LM IB

Following the upheavals at the end of MM II new categories of evidence become available and in these we see further martial flavours of Cretan Bronze Age culture in MM III. As noted above the schism between MM II and MM III is of particular social importance though it is clear from the surviving weaponry that in martial terms there is a strong degree of continuity. Evely (1996) marshalled much of the direct evidence relating to the possibility of Neopalatial warfare, and a further contextualisation of this evidence demonstrates the clear martial aspect to this period. We must exclude the valuable insights of burials as we still do not know how the overwhelming majority of the deceased were disposed of until very late in LM I. As with MM II, sites of a ritual character remain important sources of weapons finds particularly in MM III. We begin to find many more examples of figural art in the Neopalatial period and therefore may expect to find martial images.

Starting with attitudes to violence, we may begin to look at the environment surrounding a warrior ideology. With the growing wealth of many Cretan polities, cohesive political frameworks required social domination, be it ritual, political or military, or indeed a synthesis of all three (Haysom 2010). Surrounded by seafaring states or empires commanding large military forces, the wealth of Crete is itself indicative of an ability to retard aggression from outside the island as well as between polities of the island. A point made by Pollington (1996, 167) regarding Anglo-Saxon society is salient: 'aggression was never far from the surface, since any indication that there was a reluctance to defend oneself could be interpreted as an invitation for others to attack.' International trade and exchange can be see as widespread 'advertisement' of Cretan wealth, which was certainly such an 'invitation for others to attack' and so we military force is a requisite element in social and material wealth defence strategies (see Hiller 1984; Niemeier 2004; Kilian-Dirlmeier 2000).

The training of warriors, as noted above, takes a considerable dedication of time and resources, and it depends further on a systemic preparation for war and violence – war was an expected and therefore coherently planned element of political dynamics. Such training regimes would be broadly encompassing and would invoke an ethos of competitive fitness, combat art complexity and material expressions of 'warriorhood'. The Cretan ideal of the male body illustrated on innumerable media is athletic and lithely muscular (Preziosi and Hitchcock 1999, 117), an ideal well suited to physical trials. The scenes of bull-leaping that we see at Knossos in particular (Hallager and Hallager 1995), whatever their superficial symbolism, are dangerous acts requiring skill and dexterity. While not relating to warfare these events demonstrate the celebrated existence of dangerous and potentially violent behaviour in the public arena. Two young boys depicted at Akrotiri on nearby Thera demonstrate that from a very young age, youths of high status in this region (Marinatos 1971; Preziosi and Hitchcock 1999) might be expected to learn skills for violent competitions. Evely (1996, 65) has noted that the helmets on the Boxer Rhyton from Aghia Triada are evidence of the potential ferocity of such combats in Cretan Bronze Age societies, and the context of this scene ties interpersonal combat scenes to bull-leaping, both expressions of strength and skill. Other boxing scenes (Hiller 1999) are more fragmentary, one coming from a stone vase fragment from Knossos and another from a sealing from the Temple Repositories. The possible human sacrifice at Anemospilia (Sakellarakis 1997, 154), not to mention the possible cannibalism at Knossos (Wall, Musgrave and Warren 1986) are further evidence that life and death in Bronze Age Crete included a multitude of embodiments of violence against the body.

Supporting these circumstantial cases, we have some scenes which are much more directly related to warfare. Younger and Rehak's (2008, 169) reference to 'soldiers in line march[ing] off to duty' is symptomatic of the tendency amongst Aegean scholars to use euphemisms wherever possible for warfare, violence or combat. Assuming that their 'duty' would sometimes involve warfare we can see that there are several sealings from Crete which depict this overtly militaristic theme of marching warriors armed with spears and figure-of-eight shields (CMS II 8.1 No. 276, No. 277, No. 278; Papadopoulos 2006; Molloy 2006). A further scene of warriors, including one marching with spear, shield and helmet with a dog at his side (CMS II 8.1 No. 236), does not provide a clear context for the character, but the shield affords it a warlike character. More directly violent scenes exist. There are examples of a scene from both Knossos and Aghia Triada of a helmeted warrior (again with accompanying dog) chasing down another man and stabbing him from behind with a sword (CMS II.6 No. 15). Also from Aghia Triada, we have a swordsman attacking a falling opponent (CMS II.6 No. 16) and from Knossos a swordsman attacking a rearing lion (CMS V S.1A No. 135).

Two seals best provenanced to Crete depict swords being used in the most direct manner possible, stabbing over an opponent's shield to deliver a killing strike (CMS VII No. 100, CMS VII No. 130). This has clear LH I parallels in Grave Circle A at Mycenae (CMS I No. 11, CMS I No. 12), and Papadopoulos's (2006) dating of these seals to LM I indicates widespread popularity of this stock image.

Iconographic images of the boar's tusk (and similar type) helmets appear on an unprovenanced double-axe and sealings from Aghia Triada, two sealings from House A at Zakros, a stone Rhyton from Knossos and a seal provenanced to 'Crete'. Shields only survive as artistic representations, appearing on glyptic art, jewellery pendants and LM I 'Alternating Style' pottery (Rehak and Younger 2001, 438).

From Crete there are also a number of other images of warriors with swords, spears and shields depicted as symbols of power, such as the Chieftain's Cup from Aghia Triada and the Master Impression from Chania. The Chieftain's Cup illustrates a man holding a sword 'reporting' to a taller man holding a staff or spear (the tip of the object is broken off). These are accompanied by three other figures carrying large hides, possibly for the manufacture of the well known figure-of-eight or tower shields (Papadoulos 2006). Regarding the volume of images with a martial aspect, the examples cited are by no means exhaustive, and it can be noted that finds from Neopalatial Crete significantly exceed the known number of contemporary images from the Mainland in terms of quantity and diversity (Krzyszkowska 2005, 139). If we exclude the finds from the few graves at the single site of Grave Circle A at Mycenae to allow for the accident of survival, the Cretan majority is increased by several multiples (Papadopoulos 2006).

The rarity of reconstructable figural scenes in fresco art of the Neopalatial period (Preziosi and Hitchcock 1999) accounts for the absence of known scenes of warfare from the palaces – the well-known Knossian frescoes are largely from ensuing the Third Palace Period. That the warriors on the Miniature Fresco from Thera may be from Crete is hinted at by the flotilla of boats that accompanies them (Mainlanders are not known to have been powerful at sea at this time), and the attribution of a Mainland identity is paradoxically built on the belief that Minoans did not engage in warfare while the Mycenaeans did (Morgan 1988). Kilian-Dirlmeier (2000) places them as native Therans.

Returning once again to the material culture of war some caveats on the nature of the archaeological record must be set out. A handful of swords do not make an army, so the existence of a few examples is not robust evidence if taken purely at face value. We must consider related evidence to better understand their importance, particularly in relation to technology of manufacture, the accident of preservation, and find context. As noted above the technological innovations of the Mallia swords represent a mature weapon smithing tradition, and by MM III our evidence is much enhanced by other finds, particularly the swords from Arkalochori cave (Hazzidakis 1912). Of particular interest is not merely the considerable number (nine) of Type A swords found here, but that their morphological similarities indicate that at least some of these were manufactured using the same wooden former to create the design for the sword. Differences in length are negligible and are easily accounted for by the >10% variable shrinkage as moulds dry (O' Faolain 2004; O' Faolain and Northover 1998). The importance here lies in the fact that we have evidence for batch production of swords and despite these artefacts being unfinished we can see that workshops had developed in Crete which could turn swords and spears out on an industrial scale (Molloy 2010), as is indicated in the later Linear B tablets from Knossos (Macdonald and Driessen 1984, 64; Ventris and Chadwick 1973, 360-361).

Depositional traditions in Crete have created a vast bias in the record. From the Bronze Age to Medieval times, the single most important category of activity which enters weapons into the archaeological record is interment with burials. Late Protopalatial and Neopalatial burials in Crete are rare, to the extent that the known examples would account for an indeterminably small percentage of the living population. The funerary record is thus virtually invisible and deprives us of a valued source of evidence common to all known societies who had weapons traditions. This bias cannot be underestimated, as the other main context where weapons played a role was on the battlefield; sites again virtually impossible to recover weapons from due to the living having taken theirs with them and those of the dead either respectfully removed or looted. We are thus left with few contexts in which we can even expect to find weapons, yet it is interesting to note one final potential source for weapon deposition – sanctuaries and shrines. The aforementioned Crystal Pommel sword and the two Type A swords from Mallia were interred in precisely such a context, and they are not alone in this.

At Psychro Cave, one of the few cave sites to have Protopalatial through Neopalatial activity, we find a number of sword models and spears interred in a cave which was to be a later focus for weapon deposition (Boardman 1961), a long tradition mirrored also at the Idean Cave. Hundreds of bronze votive swords, daggers and double axes came from Arkalochori alongside the nine Type A examples (Hazzidakis 1912; Kilian-Dirlmeier 1993; 2000; see also Haysom 2010). At peak sanctuaries the depiction of men with daggers was noted earlier, but we also find examples of miniature models of swords of Type A form known from Psychro and Arkalochori and akin to the Crystal Pommel sword from Mallia. Large quantities of these blades are known from Juktas, Karphi, Petsophas, Modhi, Kophinas, Traostalos and probably other sanctuaries (Peatfield 1989; Kyriakidis 2005; Briault 2007; various contributions in Morris and Peatfield, in preparation). We may also note the spearhead actually in use in the shrine at Anemospilia in late MM II/early MM III, though a martial connotation is perhaps unlikely.

We thus have evidence of votive (not functional) swords coming from urban shrines, peak sanctuaries and even cave sites, all very different contexts with regional idiosyncrasies but nonetheless representing a recurrent pattern (Briault 2007). Kilian-Dirlmeier (2000, 830) has suggested that it may be appropriate to consider the existence of a deity in Crete that was concerned with 'the production and use of weaponry'. Many of these objects were not tools of combat, but the later parallel of Classical sanctuaries is evidence that special votive versions of weapons were appropriate offerings along side real weapons. In Crete, the rural sanctuary of Kato Syme was to continue this ritual/martial association with several swords discovered there of LM IIIA date (Kilian-Dirlmeier 1993).

The frequent mention (e.g. Driessen 1999) of the fact that weapons are depicted and found only (or even mainly) in ritual contexts has little bearing on their martial roles, as it illustrates that they had a multiplicity of meanings. For these objects to be imbued with any symbolic meaning presupposes a recognition of the symbol as referring to a widely recognised reality, and combat is the most obvious context where swords can be given effective meaning as objects. The discovery of swords and model swords at ritual sites is particularly important, as it is widely argued that ritual ideologies were used as a mode of elite domination. The relationship between war and religion could have taken many forms with historical contingencies, but as they are perhaps the two most effective modes of social and political domination, the interplay between them is most important.

This interconnection could also relate to the individual, particularly in relation to cathartic measures in the alleviation of guilt or other psychological impacts of warfare (Molloy and Grossman 2007). Modern military psychology demonstrates that the majority of combatants suffer from varying levels of psychological stress and damage from involvement in combat, the so-called 'Gulf War Syndrome' for example, and treatment for this is undertaken by trained psychologists (Grossman 1996; 2005; Shephard 2002). The power of religion and rituals in the mitigation of mental trauma in prehistoric warfare may have served a broadly similar function and involved processes which sanctioned and justified violence by divine mandates. Thus it is plausible that offerings of weapons or symbols of weapons in religious contexts may have been part of a process of normalisation or recovery, albeit in a subliminal guise. An alternative reading may be that in the context of other finds at sanctuaries in particular, could the widely held belief that votives were supplications for healing be paralleled with offerings of model weapons as supplications for protection or even divinely solicited martial skills?

The above mentioned weapons are complemented by a host of other finds from different areas of the island and from a wide variety of contexts. From Zakros come three Type A swords with dramatically thickened tangs which were an innovation to increase their combat durability. From Skalis comes the point of a sword which Kilian-Dirlmeier attributes to Type A, and two further examples are known from the 'isolated deposit' at the tombs of Isopata, indicating that a destroyed tomb nearby was no later than LM I in date (Evans 1935, 846; Driessen and MacDonald 1997, 71). The possibility that LM I burials with weapons occurred here is supported by recent discoveries at Poros, near the coast in the Knossos area, where burials with weapons and boars tusk helmets are known, though surviving in fragmentary form (Dimopoulou 1999). As will be argued below, after Whitley (2002), weapons in burials do not indicate warrior identity, but it is interesting to find this tradition taking place in Crete before the collapse of the Neopalatial political network.

In the Neopalatial period the palatial centres are often believed to relate to levels of power and influence based on size and wealth. It is difficult however to use such criteria (e.g. Knossos being twice the size of Mallia) to argue for political hegemony over several centuries, as this disregards historical contingencies and assumes a long period of stability with little change. As Schoep has demonstrated (2002) the case study of the Mesara shows subtle but important shifts in power bases through time, even on a regional scale. It must also be noted that political struggles or wars operate at a chronological scale much finer than the parameters of our ceramic-led chronological markers, and that wholesale destructions are a poor indicator of what might be regarded as more typical conflict resolution mechanisms. Though settlement dynamics and material culture forms are strong markers of regional power relationships, by their nature they cannot tell us about major events such as wars which influenced these broader trajectories, save for the rare events of urban destruction. As wars have been used as markers to punctuate written human history for thousands of years, this archaeological invisibility leads to a very different view of the trajectory of Cretan prehistory and the tools we typically employ to characterise social change.

The heterarchical model proposed by Schoep (2002; 2006) and Knappett (2008; see also Knappett and Schoep 2000; Wright 2004) in particular highlights the human element behind polities, with factions amongst the ruling elite vying for power with each other as well as the elites of other polities. If we envisage 'vying' in this context as including violence ranging from single combats to raids to open warfare, we can operate with a model that includes intra-polity conflicts as well as those operating on a inter-polity level. Social hierarchies are therefore fluid and not enshrined through the immutable power of a faceless palace but are rather dynamically manifested through the negotiation of power by groups. In this sense we do not need to envisage elites living in the palaces and can instead view elites manifesting power through centralising authority systems rather than centralised authority structures.

Minoans and Mycenaeans

The sudden appearance of the Shaft Graves on the Mainland has been accounted for in various ways (Dickinson 1984), from Evans' buccaneering princes (Evans 1930, 99) to Davis' opportunists controlling the flow of Transylvanian gold (Davis 1983). We see in these new traditions of grave style an increase in Minoanising pottery, the introduction of specialised combat weaponry, new art forms and new religious ideologies. Most of these categories of material have Cretan antecedents, in functional if not always typological terms. Much of the material (including weaponry) is often argued to have been actually manufactured in Crete (Catling 1974, 252; Hood 1980; Mylonas 1973, 419; Driessen and Macdonald 1984, 64 but see Dickinson 1997 for alternative views). However, this is not generally perceived as being typical Cretan material culture, and is said to have been made specifically for a Mainland audience, perpetuating the myth of warlike Mycenaeans liking violent scenes which the peaceful Minoans would have shunned. Methodologically, it is an unsound premise to assume that there was no audience for these themes in Crete. The wealth of these burials represents the archaeological wealth of the sealed context of royal burials, but until such a dataset becomes available in Crete we must be cautious in how we relate the variable categories of evidence. In terms of material culture content they are not specifically representative of either Mainland or Cretan traditions, and Evely (1996) has rightly warned about the dangers of comparing unequal datasets – the rich and unlooted grave group with settlement archaeological data.

The palatial societies of Crete were vastly better resourced and represent polities of larger populations than those of the Mainland, and so the martial character of the latter society at this time may be seen as a minor regional variation of the former, as reflected in the forms of material culture interred there (Dickinson 1997, 47). Militarily, there can be little doubt that the resources commanded by the Cretan palaces put them in a stronger position than their Mainland counterparts. Wright's (2008, 243) assertion that Mainlanders 'may originally have participated in raiding parties, but probably soon came to offer their services as warriors, either to control piracy or to provide security in and around the palaces', characterises the underlying tradition in Aegean studies of helpless Minoans and warrior Mycenaeans. Allowing for much greater two-way military interaction may account for the very particular nature of Cretan society by LM IIIA, where the Mainlanders were not hordes of invaders but long time participants in broader Aegean politics. Apart from the mainland, Cretan cultural influence is marked throughout the Aegean region in LM I, with the development of colonies or activity within existing communities over which they had strong cultural influences at Kos, Kea, Kythera, Thera, Kalimnos, Karpathos, Samos, Melos, Iasos, Cnidos and Miletus (Hiller 1984; Wiener 1989; 1999; Dickinson 1994; Davis 2001).

Whitley (2002) has made the important argument that the inclusion of weaponry in the expression of identity at a funeral-event, and perhaps for the afterlife, does not infer that the person was a warrior in life (see also Driessen and Schoep 1999; Smith 2009). These are a mere handful of elite persons with a gratuitous quantity of weaponry which exceeded any semblance of the martial requirements of an individual in life, so the martial symbolism has to be seen as equally hyperbolic in nature. As the quantity of the weaponry relates to symbolism more than martial reality, its primary importance in military terms is of how the variety of forms and sheer quantity reflect similar production logistics as found in Crete at this time.

The human toll of protracted wars or series of wars has been made evident in the millennia between the Bronze Age and the present, whereby the sheer impact of large scale fatalities in wars can take a generation to recover from. If wars and natural catastrophes combined are continual occurrences over many years, then the attrition factor can dramatically reshape demographics and social structures. Hood (1971, 383) originally argued that LM IB was a time of growing unrest and warfare rather than a brief but major war, and this was explored in detail by Driessen and MacDonald (1997), who charted architectural changes and shifts in settlement patterns. Protracted warfare fits the diminishing population and power/wealth/influence of settlements within their temporal framework. Just as Cretan martial traditions had reshaped Mainland culture in MH III (particularly material culture), Mainlanders' involvement in the teetering Neopalatial system was to have longer term effects. There is no reason to assume that every last Mainland warrior was involved in the Cretan wars, and so militarily both they and mainland non-combatants would have suffered considerably less. Farming, trade, craft industries and more would all have continued with less risk of interference than was occurring in Crete.

A Critical Generation c.1450 – 1400 BC[1]

Despite a lack of precise chronological parity, LM II and LH IIB represent a milestone in the power dynamics of the Aegean, with a marked decrease in the potential military power of Crete and the occupation of fewer urban centres. By LM II we begin to find inhumation graves as a more common form of dealing with the dead, representing a tradition not widely practiced for centuries. An analysis of the specific nature of burial practices goes beyond the scope of this paper, and Preston's (1999; 2004a; 2004b) arguments for the novelty of burial forms (tomb shape and material culture) exhibiting a mixture of Mainland, Cretan and new traditions are followed here. The widely held belief that burials with weapons reflect the arrival of Mainland warriors as either invaders or mercenaries reflects a biased reading of the material record based on the premise that the Bronze Age peoples of Crete were not characterised in death by the accoutrements of war. The materiality of burials can thus muddy the waters as they are a 'time capsule' created through an event following the death of an individual, and our veritable lack of evidence for their disposal and commemoration of the dead creates an important negative bias in our sources. The 'sudden' appearance of weapons in burials is a net result of the increasing visibility of burials in archaeological terms, a synthesis of 'Shaft Grave' and Neopalatial societies.

---

[1] This is taken to equate to LH IIB and LM II in Knossian terms, LM IIIA:1 for other areas of Crete, the absolute dates given are open to change in the course of ongoing chronological debates. LM II is used in the discussion to mean this period with due note that it is not represented in ceramic terms in most areas of Crete.

The increased hostilities of the century before LM II may well have drawn Mycenaeans to Crete as mercenaries, but also as political allies. The integration of elements of the Mainland elites into Cretan society in LM IB-LM II may have been a long process involving intermarriages and political manoeuvring as opposed to a brief event taking place in LM II. This might be seen in the beginning of the tradition of burials with weapons in LM I at Isopata and Poros, contemporary with the latest phase of similarly furnished burials in the Shaft Graves at Mycenae and around Pylos. These developments would have created a synergy between mainland elite practices and those of Crete, so that by LM II (and LM III) the occurrence of new burial forms need only reflect hybridised elite ideologies and modes of commemoration in death. There is as yet no evidence for the existence of Mainland urban centres which could have commanded the manpower and resources of the Cretan palaces or pose a serious threat to them at this time.

The synergy between martial equipment in Crete and the Mainland further indicates a continual flow of interaction between the two areas, including martial traditions. Of much importance in this regard are the marked changes in weaponry styles, with the adoption of the Type C and D swords on the Mainland and Crete simultaneously. Both martial systems were evolving in tandem at this time. It is likely on the basis of slightly later evidence that similar armour forms were in use (Verdelis 1977; Hood 1971), the shield forms were identical, heavy double axes were current in both areas, and the coeval developments of the Type F and H spearheads is further testimony to a common martial tradition (Höckmann 1980). We therefore have little need to see in the 'warrior burials' of Crete a Mycenaean ethnicity, as such a distinction between Minoan and Mycenaean at this time runs the risk of over-simplifying protracted political and social transformations.

Crete in the Time of the Mainland Citadels c.1400 - 1200 BC

Following this transitional period, many of the changes which were beginning to appear in LM II become more developed in LM IIIA, particularly those to burial practices. The LM IIIA occupation at Knossos is not on as large a scale as before (Hatzaki 2004) but it is at least as wealthy and powerful as any Mainland centre. The political landscape of the rest of Crete is dramatically different to that of the Mainland, and its historical trajectory diverges also with a diminution of centralised authorities and urban systems concurrent to the very zenith of the Mainland urban centres. The growth of Mycenaean power and influence need not be seen as occurring strictly at the expense of Crete, as Knossos still thrives and there is little reason to believe that the power of Crete now lay in the hands of specific Mainland polities. On the mainland, this period marks the development of the well-known Mycenaean citadels and palaces of the Argolid, Pylos, Athens and Boeotia. In marked contrast to these developments, there is no evidence to suggest a move to fortify the sole surviving palace at Knossos, nor is there the density of high status sites that we find in the Mycenaean heartlands.

The nature of the iconography of this period makes it difficult to ascertain much about military organisation, but finds of weaponry are still very revealing, as are the Linear B archives. The Type C and D swords are by LM IIIA1 the dominant forms, though by LM IIIA2 the first examples of Type F and G swords occur and are precursors to the large scale changes in warfare which take place in LH IIIB (Molloy in preparation). The early development of these forms in Crete indicates that in battlefield terms, the military tradition of the Cretans was still at the forefront technologically. From the Knossos tablets we have inventories of weaponry, and the numbers on even a single tablet are alarmingly high given the quantity of actual weapons that have thus far been recovered. The Ra series includes the tablet Ra1540 which lists fifty swords, while Ra 7498 lists some eighteen *and* ninety-nine, and a further twenty fragmentary tablets list an uncertain number of swords (Driessen and Macdonald 1984, 64). Ra1548 records 3 swords fitted with bindings and R 0481 lists 42 spears with bronze points, apparently including the shaft (Ventris and Chadwick 1973, 360-361). There are also more recently discovered listings of javelins, up to 200 pieces on one tablet – Vn 1341 (Shelmerdine 1999, 403). Helmets are also listed, and add yet more possible types to the already varied repertoire (Ventris and Chadwick 1956). It is clear that the weapons smiths at Knossos were producing weapons *en masse* in LM IIIA as postulated for MM II onwards, and that the examples from the burial record represent a mere fraction of the numbers in use.

The Linear B ideograms of corslets are complemented by a stone vase in the shape of a corslet with pauldrons, indicating that armour similar to that found at Dendra was also to be found in Crete at this time (Hood 1980). Recent finds of burials with weapons from Chania (Whitley 2005) indicate that martial control of the island may not have been solely based at Knossos, the largest and wealthiest settlement of this time. The dearth of settlement competition with Knossos may be indicative of its martial might, but also indicates that the real focus of power in the Aegean had shifted to the many centres of the mainland. It is possible that the wars and hardships of later LM IB had decreased the population of the island so that regional competition against the centralising influence of Knossos was undesirable.

By LM IIIB Crete was a starkly different place than it had been in the preceding centuries, with a few short-lived high-status sites at Kommos, Agia Triada and Chania, though none on a scale comparable to the contemporary Mainland centres. For a land that had the agricultural, raw material and craft resources sufficient to support several large palatial centres to become thus impoverished is somewhat surprising. During LM IIIB Crete begins to fade into obscurity in relation to Aegean power dynamics, and there is little to suggest that it featured much in the new wave of wars that ravaged the Mainland by the end of LM IIIB. By the time we find centralised settlements again in Crete in LM IIIC, they are of a scale so small and inconsistent with the palatial periods as to suggest considerable depopulation of the island as a whole. The idiosyncrasies of the Cretan weaponry (Molloy 2006), however, suggests that there had been a continuous martial tradition developing throughout the island (the eastern part in particular) which continued to be influenced to a strong degree by developments on the Mainland, not least the adoption of the Naue ii sword (Molloy 2005).

Conclusion

During the Protopalatial period of Crete, a significant shift takes place in the control of legitimate violence, through a centralisation of the craft traditions used to create weapons and new martial traditions required to use them. The ubiquitous Prepalatial and earlier Protopalatial dagger was a symbol of male identity, but by the end of the period its use was circumscribed by martial innovations which provided more specialised weaponry in the form of swords and larger shields of tower or figure-of-eight form. Swords and spears appear at sanctuary sites in MM II and models of swords are found even more frequently by MM III. There are possible defensive considerations manifested in architecture, and special buildings (Guard Houses) are built to control access in certain areas of the landscape. The most obvious 'smoking gun in hand' which can be expected archaeologically is the sacking of settlements as a result of extreme levels of warfare and violence, and such evidence is widespread at the end of MM II, along with a move in some areas to defensible sites. Given the ephemeral nature of many categories of evidence for this period, evidence for the tools of war, the results of war, and social strategies to control war are all nonetheless present by the end of the Protopalatial period.

The Shaft Graves at Mycenae and burials around Pylos represent the interment of a handful of elite individuals over the course of a century or more. They do not in any realistic sense represent the military capacities of Early Mycenaean states, but rather represent an elite ideology which celebrates martial values. Many elements used in these burials, quantities aside, were borrowed directly from the Cretan iconographic, craft and martial traditions. In itself this transference of symbolism from Crete to the Mainland is a stark warning against polarising identities and by extension martial systems. On the basis of the character of the material record, the technological capacities for weapon production, and the celebration of martial themes in art, there is nothing to differentiate the conduct of warfare in Neopalatial Crete and on the Mainland at this time.

Given the lack of a time-capsule in Crete akin to the Mainland Shaft Graves, it is remarkable how much evidence for militarism exists and that this evidence comes from a much greater variety of contexts than we find on the Mainland. It thus appears to pervade deeper into Cretan society than simply symbolism in burial practice, it features in the secular and divine mechanisms of control and would have been a key element in elite social dynamics.

Even in periods of low-intensity warfare, engaging in combat on behalf of the polity or perhaps in more localised disputes (including intra-polity disputes) was important in demonstrating the circumscribed legitimacy of violence as part of an elite mechanism which included other circumscribed activities of cultic and economic natures. By LM IB however this model can no longer be sustained, as it is clear that the increasing assaults on urban centres represent a shift in warfare and conflict resolution mechanisms. The belief in the arrival of Mainlanders to Crete in LM II reflects an unnecessary event-based perspective which a model of longer-term elite and martial interactions might better account for. Seeing an increased martial aspect to Crete in LM II/IIIA on the basis of biases in burial visibility and perhaps our ability to translate the Linear B script has at best tenuous methodological validity. There is no increase in categories of evidence that can be interpreted as directly relating to military organisation or modes of warfare. We thus have no reason to think that the increased visibility of elements of Mainland culture relates to increased warfare or visibility of martial elements in the living society. Martial interactions between Crete and the Mainland began with wholesale Cretan influence on Mainland military systems in the Shaft Graves and continued with each region perpetually influencing the other until the very end of the Bronze Age. For much of this epoch, the palatial centres of Crete were the driving force behind martial developments, their influence only ending in the decades before warfare and disasters eventually tore Aegean society apart in the years around 1200 BC.

That warfare played a structural role in political and social dynamics in prehistoric Crete is to be expected in terms of comparative data from other cultures, and is physically attested in a multitude of ways through material culture and iconography. The security to accumulate wealth, the influence of Cretans at sea and overseas, the shifting territorial boundaries, the craftsmanship of weaponry and even the character of religion all had at various times martial components. At the cusp of many changes in our periodisation of Bronze Age society in Crete, we find evidence of destructions and social change. Emphasising martial aspects to these archaeologically visible phenomena allows them to be integrated into broader social momentums as opposed to brief instances of instability or turmoil. A strong military system required training, technology, logistics, personnel, tradition and many other aspects which are created and codified in times of peace. These ensured preparedness for the inevitable conflicts which occur in densely occupied lands with wealthy regional power centres (Keeley 1996). The martial systems of Bronze Age Crete were therefore the parameters of power which allowed the accumulation of wealth in a stable social and political environment. War in Bronze Age Crete was not a collapse of order, it was a materially ordered part of society.

Acknowledgements

My thanks to Stephen and Dan for organising this conference and proceedings, and for the invitation to participate. This research was conducted under funding from the Irish Research Council for Humanities and Social Sciences. Alan Peatfield, Angelos Papadopoulos and Ioannis Georganas have provided many stimulating discussions and ideas for this research.

Bibliography

Alexiou, S. 1979. Τειχη και αχροπολεισ στη μινωιχη Κρητη (ο μυθοσ τησ μινωιχη σ ειρηνησ). *Kretalogia* 8, 41-56.

Alusik, T. 2007. *Defensive Architecture of Prehistoric Crete.* Oxford, BAR Publishing.

Avila, R. A. J. 1983. *Bronzene Lanzen- und Pfeilspitzen der griechischen Spätbronzezeit* (Prähistorische Bronzefunde V,I). Munich, Beck.

Bintliff, J. L. 1984. Structuralism and Myth in Minoan Studies. *Antiquity* 58, 33-387.

Boardman, J. 1961. *The Cretan collection in Oxford: the Dictaean Cave and Iron Age Crete.* Oxford, Clarendon Press.

Branigan, K. 1974. *Aegean Metalwork of the Early and Middle Bronze Age.* Oxford, Clarendon Press.

Branigan, K. 1999. The Nature of Warfare in the Southern Aegean During the Third Millennium BC. In R. Laffineur (ed.), *POLEMOS: le contexte guerrier en Egée à l'âge du Bronze* (Aegaeum 19), 87-96. Liège, Université de Liège.

Briault, C. 2007. Making Mountains Out of Molehills in the Bronze Age Aegean: Visibility, Ritual Kits and the Idea of a Peak Sanctuary. *World Archaeology* 39(1), 122-141.

Bridgford, S. 1997. Mightier than the Pen? (An Edgewise Look at Irish Bronze Age Swords). In J. Carmen (ed.), *Material Harm: Archaeological Studies of Warfare and Violence,* 95-115. Glasgow, Cruithne Press.

Bridgford, S. 2000. *Weapons Warfare and Society in Britain: 1250-750 BC,* Unpublished PhD thesis, University of Sheffield.

Burke, B. 2005. Materialisation of Mycenaean Ideology and the Ayia Triada Sarcophagus. *American Journal of Archaeology* 109, 403-422.

Callaghan, P. J. 1994. Archaic, Classical and Hellenistic Knossos – A Historical Summary. In D. Evely, H. Hughes-Brock and N. Momigliano (eds.), *Knossos: A Labyrinth of History – Papers Presented in Honour of Sinclair Hood,* 135-141. Athens, British School at Athens.

Chapouthier, F. 1938. *Deux épées d'apparat découvertes en 1936 au palais de Mallia au cours des fouilles executes au nom de l'École française d'Athènes* (Études crétoises 5). Paris, P. Geuthner.

Chryssoulaki, S. 1999. Minoan Roads and Guard Houses – War Regained. In R. Laffineur (ed.), *POLEMOS: le contexte guerrier en Egée à l'âge du Bronze* (Aegaeum 19), 75-86. Liège, Université de Liège.

Clements, J. 2007. The Myth of Thrusting Versus Cutting Swords. In B. P. C. Molloy (ed.), *The Cutting Edge: Archaeological Studies in Combat and Violence,* 168-177. Stroud, Tempus.

Davis, E. 1983. The Gold of the Shaft Graves: the Transylvanian Connection. *Temple University Aegean Symposium* 8, 32-38.

Davis, J. L. 2001. The islands of the Aegean. In T. Cullen (ed.), *Aegean Prehistory: A Review,* 19-76. Boston, Archaeological Institute of America.

Dickinson, O. T. P. K. 1984. Cretan Contact with the Mainland in the Time of the Shaft Graves. In R. Hägg and N. Marinatos (eds.), *The Minoan Thalassocracy: Myth and Reality,* 115-117. Stockholm, Svenska Institutet i Athen.

Dickinson, O. T. P. K. 1997. Arts and Artefacts in the Shaft Graves: Some Observations. In R. Laffineur and P. P. Betancourt (eds.), *TEXNH: Craftsmen, Craftswomen and Craftsmanship in the Aegean Bronze Age* (Aegaeum 16), 47-49. Liège, Université de Liège.

Dimopoulou, N. 1999. The Neo-Palatial Cemetery of the Knossian Harbour-Town at Poros: Mortuary Behaviour and Social Ranking. In *Eliten in der Bronzezeit: Ergebniesse zweier Kolloquien in Mainz und Athen,* 27-36. Mainz, Verlag des Römisch-Germanischen Zentralmuseums.

Driessen, J. 1999. The Archaeology of Aegean Warfare. In R. Laffineur (ed.), *POLEMOS: le contexte guerrier en Egée à l'âge du Bronze* (Aegaeum 19), 11-21. Liège, Université de Liège.

Driessen, J. and Macdonald, C. 1984. Some Military Aspects of the Aegean in the Late Fifteenth and Early Fourteenth Centuries BC. *Annual of the British School at Athens* 79, 49-75.

Driessen, J. and Macdonald, C. 1997. *The Troubled Island: Minoan Crete Before and After the Santorini Eruption* (Aegaeum 17). Liège, Université de Liège.

Driessen, J. and Schoep, I. 1999 The Stylus and the Sword — The Role of Warriors and Scribes in the

Conquest of Crete in R. Laffineur (ed.) *POLEMOS: le contexte guerrier en Egée à l'âge du Bronze* (Aegaeum 19), 385-392. Liège, Université de Liège.

van Effenterre, H. 1980. *Le palais de Malia et la cité Minoenne* Vol. I. Rome, Edizione dell' Ateno.

Evely, D. 1996. The Neo-Palatial Warrior: Fact or Fiction,. In D. Evely, I. Lemos and S. Sherratt (eds.), *Minotaur and Centaur: Studies in the Archaeology of Crete and Euboea Presented to Mervyn Popham*, 59-69. Oxford, BAR Publishing.

Evans, A. 1930. *The Palace of Minos at Knossos III*. London, Macmillan.

Evans, A. 1935. *The Palace of Minos at Knossos IV*. London, Macmillan.

Ó Faoláin, S. 2004. *Bronze Artefact Production in Late Bronze Age Ireland: A Survey*. Oxford, BAR Publishing.

Ó Faoláin, S. and Northover, J. P. 1998. The Technology of Late Bronze Age Sword Production in Ireland. *The Journal of Irish Archaeology* 9, 69-88.

Grossman, D. 1996. *On Killing: The Psychological Cost of Learning to Kill in War and Society*. New York, Back Bay Books.

Grossman, D., with Christensen, L. W. 2004. *On Combat: The Psychology and Physiology of Deadly Conflict in War and Peace*. New York, PPCT Research Publications.

Haggis, D. 2007. Stylistic Diversity and Diacritical Feasting at Protopalatial Petras: A Preliminary Analysis of the Lakkos Deposit. *American Journal of Archaeology* 111, 715-775.

Hallager, B. P. and Hallager E. 1995. The Knossian Bull – Political Propaganda in Neopalatial Crete. In R. Laffineur and W. D. Niemeier (eds.), *POLITEIA: Society and State in the Aegean Bronze Age* (Aegaeum 19), 547-556. Liège, Université de Liège.

Hamilakis, Y. 2002. Too Many Chiefs? Factional Competition in Neopalatial Crete. In J. Driessen, I. Schoep, and R. Laffineur (eds.), *Monuments of Minos: Rethinking the Minoan Palaces* (Aegaeum 23), 179-199. Liège, Université de Liège.

Hatzaki, E. 2004. From Final Palatial to Postpalatial Knossos: A View from the Late Minoan II to Late Minoan IIIB Town. In G. Cadogan, E. Hatzaki and A. Vasalakis (eds.), *Knossos: Palace, City, State*, 121-126 London, British School at Athens.

Hatzi-Vallianou, D. 1979. Κεντρική Κρήτη, Νομός Ηρακλείου. *Archaiologikon Deltion* 134 (β2), 382-385.

Haysom, M. 2010. The Double-Axe: A Contextual Approach to the Understanding of a Cretan Symbol in the Neopalatial Period. *Oxford Journal of Archaeology* 29(1), 35-55.

Hazzidakis, J. 1912. A Minoan Sacred Cave. *Annual of the British School at Athens* 19, 35-47.

Hiller, S. 1984. Pax Minoica versus Minoan Thalassocracy. Military aspects of Minoan Culture. In R. Hägg and N. Marinatos (eds.), *The Minoan Thalassocracy: Myth and Reality*, 27-30. Stockholm, Svenska Institutet i Athen.

Höckmann, O. 1980. Lanze und Speer im spatminoischen und mykenischen Griechenland. *Jarbuch des Römisch-germanischen Zentralmuseums, Mainz* 27, 13-158.

Höckmann, O. 1999. Scenes of Warfare and Combat in the Arts of the Aegean Late Bronze Age. Reflections on Typology and Development. In R. Laffineur (ed.), *POLEMOS: le contexte guerrier en Egée à l'âge du Bronze* (Aegaeum 19), 319-331. Liège, Université de Liège.

Hood, M. S. F. 1971. *The Minoans: Crete in the Bronze Age*. London, Thames and Hudson.

Hood, M. S. F. 1980. Shaft Grave Swords: Mycenaean or Minoan? In *Acts of the Fourth International Cretological Congress*, 223-241. Athens.

Keeley, L. H. 1996. *War Before Civilization: The Myth of the Peaceful Savage*. Oxford, Oxford University Press.

Keegan, J. 1994. *A History of Warfare*. London, Pimlico.

Kilian-Dirlmeier, I. 1993. *Die Schwerter in Griechenland (außerhalb der Peloponnes), Bulgarien und Albanien* (Prähistorische Bronzefunde IV,12). Stuttgart, Franz Steiner Verlag.

Kilian-Dirlmeier, I. 2000. Thera and Warfare. In S. Sherratt (ed.), *The Wall Paintings of Thera: Proceedings of the First International Symposium*, 825-830. Athens, Thera Foundation.

Knappet, C. 2008. The Protopalatial Period: The Material Culture. In C. W. Shelmerdine (ed.), *The Cambridge Companion to the Aegean Bronze Age*, 121-140. Cambridge, Cambridge University Press.

Knappet, C. and Schoep, I. 2000. Continuity and Change in Minoan Political Power. *Antiquity* 74, 365-371.

Kristiansen, K. 2002. The Tale of the Sword – Swords and Swordfighters in Bronze Age Europe. *Oxford Journal of Archaeology* 21(4), 319-332.

Krzyszkowska, O. 2005. *Aegean Seals: An Introduction*. London, Institute of Classical Studies.

Kyriakidis, E. 2000. A Sword Type on the Chieftain's Cup. *Kadmos* 39, 79-82.

Kyriakidis, E. 2005. *Ritual in the Bronze Age Aegean: The Minoan Peak Sanctuaries*. London, Duckworth.

MacGillivray, J. A. 2000. *Minotaur: Sir Arthur Evans and the Archaeology of the Minoan Myth*. London: Jonathan Cape.

Marinatos, S. 1971. *Excavations at Thera IV*. Athens, Archaeological Society of Athens.

Molloy, B. P. C. 2005. The Adoption of the Naue ii Sword in the Aegean. In C. Briault, J. Green, A. Kaldelis and A. Stellatou (eds.), *Proceedings of SOMA 2003*, 115-117. Oxford, BAR Publishing.

Molloy, B. P. C. 2006. *The role of Combat Weaponry in Bronze Age Societies: The Cases of the Aegean and Ireland in the Middle and Late Bronze Age*. Unpublished PhD thesis, University College Dublin.

Molloy, B. P. C. 2008. Martial Arts and Materiality: A Combat Archaeology Perspective on Aegean Swords of the Fifteenth and Fourteenth Centuries BC. *World Archaeology* 40(1), 116-134.

Molloy, B. P. C. 2010. Swords and Swordsmanship in the Aegean Bronze Age. *American Journal of Archaeology* 114(3), 403-428.

Molloy, B. P. C. and Grossman, D. 2007. Why Can't Johnny Kill? The Psychology and Physiology of Interpersonal Combat in B. P. C. Molloy (ed.), *The Cutting Edge: Studies in Ancient and Medieval Combat*, 188-202. Stroud, The History Press LTD.

Morgan, L. 1988. *The Miniature Wall Paintings from Thera: A Study in Aegean Culture and Iconography*. Cambridge, Cambridge University Press.

Mylonas, G. 1973. *Ο ταφικός κύκλος Β των Μυκηνών*. Athens.

Niemeier, W.-D. 1983. The Character of the Knossian Palace Society in the Second Half of the Fifteenth Century BC: Mycenaean or Minoan? In O. Krzyszkowska and L. Nixon (eds.), *Minoan Society*, 217-236. Bristol, Bristol Classical Press.

Niemeier, W.-D. 2004. When Minos Ruled the Waves: Knossian Power Overseas. In G. Cadogan, E. Hatzaki, and A. Vasilakis, (eds.), *Knossos: Palace, City, State*, 393-398. London, British School at Athens.

Nixon, L. 1983. Changing Views of Minoan Society. In O. Krzyszkowska and L. Nixon (eds.), *Minoan Society*, 237-245. Bristol, Bristol Classical Press.

Nowicki, K. 2008. *Monastriaki Katalimata: Excavation of a Cretan Refuge Site 1999-2000*. Philadelphia, Institute for Aegean Prehistory.

Oakeshott, E. 1999. *The Archaeology of Weapons: Arms and Armour from Prehistory to the Age of Chivalry*. Woodbridge, Boydell.

Papadopoulos, A. 2006. *The Iconography of Warfare in the Bronze Age Aegean*. Unpublished Ph.D. thesis, University of Liverpool.

Peatfield, A. A. D. 1989. *The Peak Sanctuaries of Minoan Crete*. Unpublished PhD thesis, University College London.

Peatfield, A. A. D. 1990. Minoan Peak Sanctuaries: History and Society. *Opuscula Atheniensia* 18, 117-131.

Peatfield, A. A. D. 1999. The Paradox of Violence: Weaponry and Martial Art in Minoan Crete. In R. Laffineur (ed.), *POLEMOS: le contexte guerrier en Egée à l'âge du Bronze* (Aegaeum 19), 67-74. Liège, Université de Liège.

Peatfield, A. A. D. 2007. Reliving Greek Personal Combat: Boxing and Pankration. In B. P. C. Molloy (ed.), *The Cutting Edge: Studies in Ancient and Medieval Combat*, 20-34. Stroud, The History Press LTD.

Peatfield, A. A. D. and C. Morris. In preparation. *Proceedings of the Peak Sanctuary Workshop held in Dublin, February 2010*.

Pollington, S. 1996. *The English Warrior from Earliest Times to 1066*. Norwich, Anglo-Saxon Books.

Popham, M. R. and Catling, H. W. 1974. Sellopoulo Tombs 3 and 4, Two Late Minoan Graves Near Knossos. *Annual of the British School at Athens* 69, 195-258.

Preston, L. 1999. Mortuary Practices and the Negotiation of Social Identities at LM II Knossos. *Annual of the British School at Athens* 94, 131-143.

Preston, L. 2004a. A Mortuary Perspective on Late Minoan Crete. *American Journal of Archaeology* 108, 321-349.

Preston, L. 2004b. Final Palatial Knossos: A Mortuary Perspective on Political Dynamics. In G. Cadogan, E. Hatzaki and A. Vasilakis (eds.), *Knossos: Palace, City, State*, 137-147. London, British School at Athens.

Preziosi, D. and Hitchcock, L. A. 1999. *Aegean Art and Architecture*. Oxford, Oxford University Press.

Rehak, P. and Younger, J. G. 2001. Neopalatial, Final Palatial and Postpalatial Crete. In T. Cullen (ed.) *Aegean Prehistory: A Review*, 383-465. Boston, Archaeological Institute of America.

Sakellarakis, Y. and Sakellarakis, E. 1997. *Archanes: Minoan Crete in a New Light*. Athens, Eleni Nakou Foundation.

Sandars, N. K. 1961. The First Aegean Swords and Their Ancestry. *American Journal of Archaeology* 65, 17-29.

Schoep, I. 2006. Looking Beyond the First Palaces: Elites and the Agency of Power in EM III-MM II Crete. *American Journal of Archaeology* 110, 37-64.

Shelmerdine, C. W. 1999. Pylian Polemics: The Latest Evidence on Military Matters. In R. Laffineur (ed.), *POLEMOS: le contexte guerrier en Egée à l'âge du Bronze* (Aegaeum 19), 403-411. Liège, Université de Liège.

Shelmerdine, C. W. (ed.). 2008. *The Cambridge Companion to the Aegean Bronze Age*. Cambridge, Cambridge University Press.

Smith, S. K. 2009. Skeletal Evidence for Militarism in Mycenaean Athens. In L. A. Schepartz, S. C. Fox and C. Bourbou (eds.), *New Directions in the Skeletal Biology of Greece* (Hesperia Supplement 43), 99-109. Athens, American School of Classical Studies at Athens.

Starr, C. G. 1984. Minoan Flower Lovers. In R. Hägg and N. Marinatos (eds.), *The Minoan Thalassocracy: Myth and Reality*, 1-12. Stockholm, Svenska Institutet i Athen.

Shephard, B. 2002. *A War of Nerves: Soldiers and Psychiatrists 1914-1994*. London, Pimlico.

Tsipopoulou, M. 1999. From Local Centre to Palace: The Role of Fortification in the Economic Transformation of the Siteia Bay Area, East Crete. In R. Laffineur (ed.), *POLEMOS: le contexte guerrier en Egée à l'âge du Bronze* (Aegaeum 19), 179-189. Liège, Université de Liège.

Tzedakis, Y., Chryssoulaki, S., Venieri, Y. and Avgouli, M. 1990. Les Routes Minoennes: Rapport Preliminaire – Defense de la circulation ou circulation de la defense? *Bulletin de correspondance hellénique* 113, 43-75.

Ventris, M. and Chadwick, J. 1956. *Documents in Mycenaean Greek*. Cambridge, Cambridge University Press.

Verdelis, N. M. 1977. The Metal Finds. In P. Åström, *The Cuirass Tomb and Other Finds at Dendra*, 28-65. Goteborg, Paul Åströms Verlag.

Wall, S. M., Musgrave J. H. and Warren, P. M. 1986. Human Bones from a Late Minoan IB House at Knossos. *Annual of the British School at Athens* 81, 333-388.

Watrous, L. V. 2001. Crete from Earliest Prehistory Through the Protopalatial Period. In T. Cullen (ed.), *Aegean Prehistory: A Review*, 157-224. Boston, Archaeological Institute of America.

Wedde, M. 2000. *Towards a Hermeneutics of Aegean Bronze Age Ship Imagery* (Peleus Studien zur Archäologie und Geschichte Griechenlands und Zyperns 6). Mannheim, Bibliopolis.

Wiener, M. H. 1989. The Isles of Crete? The Minoan Thalassocracy Revisited. In D. A. Hardy, C. G. Doumas, J. A. Sakellarakis and P. M. Warren (eds.), *Thera and the Aegean World III* Volume I, 128-161. London, Thera Foundation.

Wiener, M. H. 1999. Present Arms/Oars/Ingots. In R. Laffineur (ed.), *POLEMOS: le contexte guerrier en Egée à l'âge du Bronze* (Aegaeum 19), 411-423. Liège, Université de Liège.

Whitley, J. 2002. Objects With Attitude: Biographical Facts and Fallacies in the Study of Late Bronze Age and Early Iron Age Warrior Graves. *Cambridge Journal of Archaeology* 12(2), 217-232.

Whitley, J. 2003. Archaeology in Greece 2002-2003. *Archaeological Reports* 49, 1-88.

Whitley, J. 2005. Archaeology in Greece 2004-2005. *Archaeological Reports* 51, 1-118.

Wright, J. 2004. The Emergence of Leadership and the Rise of Civilization in the Aegean. In J. C. Barrett and P. Halstead (eds.), *The Emergence of Civilisation Revisited*, 64-89. Oxford, Oxbow Books.

Wright, J. 2008. Early Mycenaean Greece. In C. W. Shelmerdine (ed.), *The Cambridge Companion to the Aegean Bronze Age*, 230-258. Cambridge, Cambridge University Press.

Younger, J. G. 1988. *The Iconography of Late Minoan and Mycenaean Sealstones and Finger Rings*. Bristol, Bristol Classical Press.

Younger, J. G. and Rehak, P. 2008. Minoan Culture: Religion, Burial Customs and Administration. In C. W. Shelmerdine (ed.), *The Cambridge Companion to the Aegean Bronze Age*, 165-186. Cambridge, Cambridge University Press.

Icon of Propaganda and Lethal Weapon: Further Remarks on the Late Bronze Age Sickle Sword

Carola Vogel

Abstract

Among the multitude of weapons in the tomb of Tutankhamun two different shaped sickle swords have been discovered. They belong to a small group of Late Bronze Age scimitars known from various places in the Ancient Near East, now dispersed between a number of museums worldwide.

25 years ago, Hans Wolfgang Müller compiled all sickle swords that were known to him with the aim of establishing a proper typology and creating a chronological framework for this weapon. His efforts resulted in a monographical study that has become a standard reference ever since: Der Waffenfund von Balata-Sichem und Die Sichelschwerter *(1987). Apart from this painstaking research the curved sword is discussed in depth in works, which deal with the weapon as an icon of divine power and instrument for ceremonial execution. Since Müller's compilation new swords have come to light that demand a re-examination of the whole group. Based on such an updated corpus, the aim of this paper is twofold. Firstly, an overview of royal iconography will exemplify the prominent role that the curved sword played in New Kingdom propaganda. A second approach will deal with the use of the sickle sword as a standard piece of equipment for infantry and chariot crews by discussing its archaeological, epigraphic and iconographic evidence. Furthermore, the practical value of this weapon in close combat will be highlighted in order to show its increasing significance from the Second Intermediate Period onwards.*

Introduction

In the following I would like to discuss some special aspects of the curved bronze sword that has found broader acceptance as the so-called 'sickle sword'. One should be aware that this name refers to its shape and not to its function, as the cutting edge lies on its outer rim.

The Egyptians themselves – devoted to equivocal expressions – named the weapon *khepesh* due to the resemblance of the curved blade to the foreleg of a cow/ox and its religious and political connotations beyond. Thus, the idea of being equipped with a 'mighty arm' implies being powerful in combat as well as an assumed general victory power.[1]

Before I discuss the two main facets of the Late Bronze Age sickle sword as an icon of propaganda and as a lethal weapon, I would like to give a brief overview of the research already conducted with respect to this weapon.

Research History

A short article published by Lenk-Chevitch in 1941 (Lenk-Chevitch 1941) might be taken as the first hint at a forthcoming academic interest in this type of sword. Despite the fact that the map Lenk-Chevitch presented to illustrate the distribution of the sickle sword appears roughly sketched in an age of vector graphics one should appreciate the author for drawing attention to it (Figure 1).

Hans Wolfgang Müller's monograph *Der Waffenfund von Balata-Sichem und Die Sichelschwerter* (1987) compiled all sickle swords that were known to publication date, resulting in a work that has became a standard reference. More recent input arrives from a dissertation by Javier Martínez Babón (Martínez Babón 1995),[2] and the typological research of Shalev (2004),[3] and Vogel (2006). Beyond these works, information about this type of weapon usually comes from short excavation reports or catalogue entries. The sickle sword is also discussed in depth in all those works which deal with the weapon as an icon of divine power and instrument for ceremonial execution. In particular *Wirkmächtige Siegeszeichen* by Othmar Keel (1974), the dissertation of Sylvia Schoske (1995), *The Pharaoh Smites His Enemies* by Emma Swan Hall (1986), and *Ceremonial Execution and Public Rewards* by Alan Schulman (1988).

However, since the fundamental monographs of Müller and Martínez Babón some new swords have come to light, demanding a re-examination of the whole group. Based on an updated corpus, the aims of this paper are twofold:

- An overview of royal iconography exemplifying the prominent role which the sickle sword played in New Kingdom propaganda.
- The use of the sickle sword as a piece of standard infantry equipment by discussing its archaeological, epigraphic and iconographic evidence. Furthermore, the practical value of

---

[1] For an in-depth discussion of the term *ḫpš*, see Martínez Babón (1995, 56-73), and Galán (1995, 69-73). For its interpretation as 'Waffenfähigkeit' I refer to Gundlach (2009, 59-65).

[2] My thanks are due to the author who provided me with a copy of his dissertation, which is currently available only in microfiche format.
[3] I owe thanks to Dr Yossi Mizrachi, Haifa, for drawing my attention to this title.

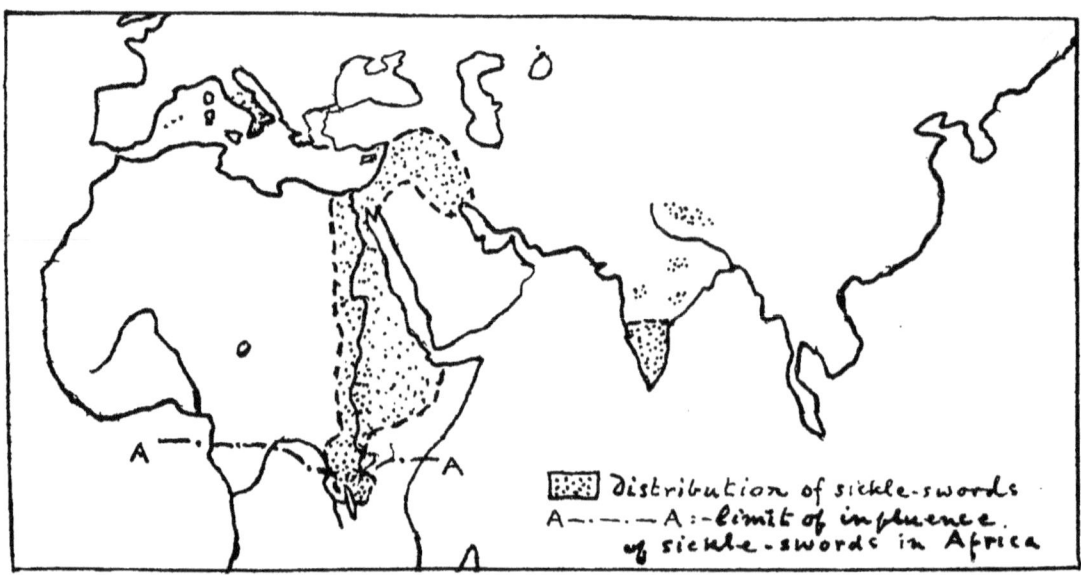

Figure 1: Distribution of the sickle sword after Lenk-Chevitch (1941, 83, fig. 43).

this weapon in close combat will be highlighted to show its growing importance beginning from the Middle Kingdom/Second Intermediate Period (MB II) onwards.

1 Icon of Propaganda[4]

1.1 The king smites the enemy with the sickle sword in absence of a god

A familiar icon of propaganda is represented by the famous group which simply shows the king smiting the enemy in the absence of a god. Examples vary from small scarabs to monumental wall paintings. Representations sufficient to highlight this significant group are:

A well-engraved scene from a scarab of Amenhotep III currently in Berlin (SMB-PK, 32326) (Petschel and von Falck 2004, 65; Schoske 1997, 15). The king is depicted in full stride, smiting an enemy who is desperately struggling to escape on his knees (Figure 2). The king – appropriately named '*neb khepesh*' – prevents his flight by putting his foot on the enemy's right thigh. The text below the scene reads 'all countries, all foreign countries' and could be understood as 'all countries are under the pharaoh's feet'.

An exceptional representation occurs in the Amarna period, originating from Talatat blocks discovered in Hermopolis (P. C. 67, Boston, Museum of Fine Arts, BMFA 64.521, P. C. 55 New York, Norbert Schimmel Collection, on loan to the Metropolitan Museum of Art) and Luxor (Tawfik 1975, 162-163, Loeben 2008, 261, incl. Abb. 10). On the example from Hermopolis Nefertiti is shown in a cabin on the royal barque, smiting a female enemy with a sickle sword. Nefertiti has adopted the royal male iconography, and acts as her husband in a comparable scene to the right thought to complete the sequence (Figure 3), (Swan Hall 1986, 25-26, incl. fig. 39, 40).[5]

The item from Luxor shows the queen in at least three shrines (Figure 4), twice smiting a female enemy with the scimitar and trampling another woman in the form of a female sphinx (Tawfik, 1975, 162-163; Roth, 2002, 29; Loeben 2008, 261). According to Dirk Bröckelmann the four cabins are represented overlapping each other, thus – against earlier interpretations – they do not belong to a single boat but to a row of ships (Roth 2002, 27-29, esp. footnote 82).

Two stelae found in the Wadi Sanur, now in Munich (Swan Hall 1986, 31, fig. 54; Schoske 1995, 18-19; Petschel and von Falck 2004, 59-60, fig. 53) and Cairo (Swan Hall 1986, 31, fig. 53) depict Ramesses II smiting enemies. Whereas the stela from Munich shows the king acting with the mace-axe in front of the god Seth, the example from Cairo depicts the king using the sickle sword without the support of a god. As on the aforementioned scarab of Amenhotep III, Ramesses is called '*neb khepesh*', but he is here additionally named as Horus.

1.2 The king receives the sickle sword from a god

A large group of monuments shows the King receiving the sickle sword from a god. These representations can be divided into further sub-groups.

1.2.1 The king receives the sickle sword from a god in order to smite Egypt's enemies as an abbreviated icon

This kind of interaction between god and king is known

---

[4] It cannot be the aim of this paper to provide a complete list of all known scenes covering this topic, but to present an overview of the various motifs.

[5] Vomberg (2004, 150-151, incl. Abb. 72) discusses the feature of the royal cabins, defining them as 'Erscheinungsfensterkioske'.

Figure 2: Smiting scene on a scarab of Amenhotep III, Berlin SMB-PK, 32326. Drawing Andreas Vogel, on the basis of a photo published in Petschel and von Falck (2004, 65, fig. 60).

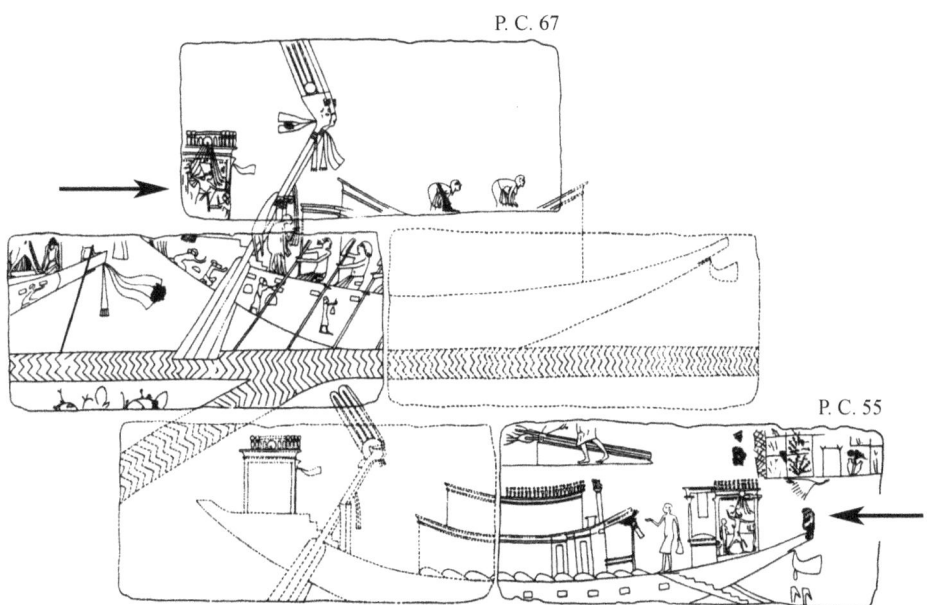

Figure 3: Nefertiti smiting a female enemy. Talatat blocks discovered in Hermopolis, P. C. 67, Boston, Museum of Fine Arts, BMFA 64.521, 63.260; P. C. 55 New York, Norbert Schimmel Collection, on loan to the Metropolitan Museum of Art. After Vomberg (2004, 314, Abb. 72).

Figure 4: Nefertiti depicted while smiting female enemies with the scimitar, and shown as a female sphinx trampling a woman. Talatat block from Luxor. After Tawfik (1975, 163, fig. 1).

on monuments from the reign of Thutmosis III onwards. Surprisingly, the weapon which the king receives from the god to smite the enemies – the sickle sword – is in most cases not used to execute the divine order. Egyptian iconography clearly distinguishes between the provision of divine power expressed by the *khepesh* sword and execution by traditional/archaic weapons of warfare. However, examples also exist where the two weapons do match, and these will be considered next.

1.2.1.1 The king receives the sickle sword from a god and smites the enemy with it

Two comparable scarabs in the British Museum show Thutmosis III receiving the sickle sword from Nefertem and from Montu (Keel 1974, 51-58). In both cases the king smites the enemy with a sickle sword (Figure 5).

A famous ivory wrist guard from Amarna shows Thutmosis IV in full stride, smiting an Asiatic enemy (Galán 1995, 72). As on many monuments he is not only equipped with the sickle sword but also with bow and arrows (Figure 6). This combination of weapons is a common feature to demonstrate that the king is well prepared for long distance and close combat warfare. Here we have the rare case that the weapon handed over by Montu is the one which Thutmosis IV actually uses.

An interesting variation appears in the Amarna-Period:

Akhenaton receives a mace and a sickle sword from Aton who is represented as the sun disc (Figure 7) (Schoske 1995, 327-328, incl. fig. a93/l-B-i-B). This scene is important as it shows quite clearly that Akhenaton felt obliged to stay within the traditional iconography, thus the king is in need of the 'divine victory power' to succeed over the enemy. The former gods such as Amun and Montu had to be replaced but the weapons remain the same: the old traditional mace, and the sickle sword which was the new icon of warfare in the New Kingdom.

Numerous examples of smiting scenes are known from Medinet Habu. On the eight columns of the south colonnade in the first court this iconography is particularly present (Nelson 1932, plates 120-122). Each of the double-scenes on their shafts show the king smiting captives before a god. The latter rewards him with the sickle sword or, less often, with a mace. In three cases we see Ramesses III receiving the scimitar from Amun-Ra and smiting the enemies with an identical weapon (Nelson 1932, pl. 120b.c, pl. 121b).

At this juncture the question occurs as to whether there are indications of chronological development. It seems that the examples in which the weapons match are dominant within the 18th Dynasty, whereas in the 19th and 20th Dynasties other weapons come to dominate. However, a conclusion would be premature.

1.2.1.2 The king receives the sickle sword from a god and smites the enemy with another weapon

The so-called 'northern rock stela' from Aswan represents Amenhotep III, who smites two enemies with the mace-axe in the presence of Khnum, Amun-Ra and Ptah, but receives the sickle sword from Amun-Ra (Swan Hall 1988, 22, incl. fig. 33, Klug 2002, 422-424, incl. Abb. 31).[6] The stela dates from the king's fifth regnal year and includes a narrative text about a rebellion in Kush which had been successfully overthrown by Amenhotep III. Gundlach favours the idea that the victory celebrated by Amenhotep III as named in this and further

---

[6] On the contrary, Klug refers to the king's weapon as a scimitar. However, a close look at the different shaped weapons represented in the scene makes it seem more reasonable to identify the king's weapon as a sickle sword.

Figure 5: Thutmosis III receives the sickle sword from Nefertem (left) and Montu (right), the same weapon he uses to smite the enemies. Scarabs, BM London. After Keel (1974, Abb. 21a, 21b).

Figure 6: Thutmosis IV in full stride, smiting an Asiatic enemy. Ivory wrist guard from Amarna. After Keel (1974, 171 Abb. 22).

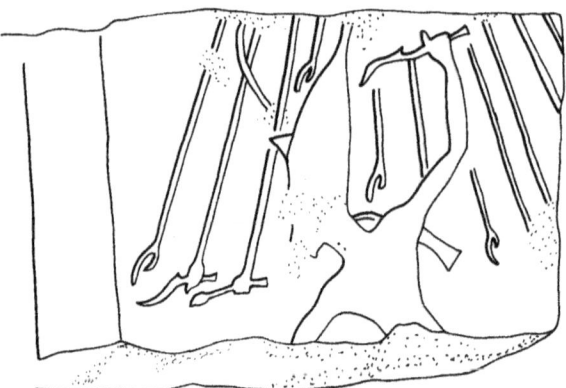

Figure 7: Akhenaton receives mace and sickle sword from Aton. Talatat-block from Karnak. After Schoske (1995, 327-328, incl. Abb. a93/l-B-i-B).

stelae from the Aswan region is fictitious (Gundlach 2004, 195-219, esp. 215).

A unique variation is depicted on a seal cylinder from Beth Shean. Ramesses II receives the sickle sword from the hybrid god Seth/Reschef/Mekal but uses a long distance weapon, the bow, to fire at a copper-ingot target beneath which two Syrian captives are bound to a stake (Figure 8) (Keel 1974, 64).

Figure 8: Seal cylinder from Beth Shean. Ramesses II receives the sickle sword from the hybrid god Seth/Reschef/Mekal. After Keel (1974, 174 fig. 27).

Rock Stela Abu Simbel 24 dedicated by the vice-king of Kush, Setau, shows Ramesses II smiting the enemy in front of Horus, Lord of Buhen, to the left and of Amun-Ra to the right. Both gods present the sickle sword to him, whereas the king kills[7] with the mace-axe (Figure 9) (Raedler 2003, 142, No. 68).

In a further representation Merenptah is smiting the enemies with the mace-axe in front of the god Ptah, who presents to him the combined sceptre of was, ankh and djet as well as the sickle sword. The scene originates from a lintel in Kom el-Qala (Figure 10) (Hornung and Staehlin 2006, 70-71).

1.2.2 Handing over of the sickle sword as the first scene within a sequence as an initial act before the campaign starts

The next group to be discussed shows the motif of handing over the sickle sword as the first scene within a sequence – indicating an initial act before the campaign starts.[8] In such cases the god gives the order to go to war. It is attested under Horemheb in Gebel es-Silsila (Heinz 2001, 240). Under Ramesses II? (Heinz 2001, 253), Merenptah (Heinz 2001, 297), and under Ramesses III in a scene at Medinet Habu which is dedicated to his first Libyan war (Figure 11) (Heinz 2001, 300). The accompanying text in the latter scene reads:

'Take the khepesh- sword, my beloved son!
Cut off the heads of the rebellious foreign countries'

In this scene the sickle sword is explicitly intended to be an icon of victory-power transferred from the god to the king.

1.2.3 Scenes in which the god rewards the king with the sword at the end of an already accomplished and naturally successful mission

Traditionally it is the god Amun or Amun-Ra who hands it over to the king.

On the northern outer wall of the great hypostyle hall at Karnak two monumental scenes show Seti I smiting the nine bows with a mace in front of Amun-Ra, who presents the sickle sword to him (Figure 12). The god leads a huge number of captured enemies (identified by their town cartouches) who were taken by his successful Horus, Seti I. The accompanying text states as follows:

šsp n=k ḫpš nsw nḫt ḥw.n ḥḏ=k pḏ.t 9

'Take the khepesh, mighty king whose mace has smitten the nine-bows'

---

[7] For a debate about the question of symbolic violence, see now Müller-Wollermann (2009).

[8] I would like to thank Dr. Marcus Müller, Potsdam, who provided me with parts of his unpublished dissertation on war reliefs (Müller 1998). I refer especially to chapter VII. 1.1 Bildtyp 1: Der König erhält vom Gott den Kriegsauftrag.

Figure 9: Ramesses II receives the sickle sword from Horus, Lord of Buhen (left), and Amun-Ra (right). Rock Stela Abu Simbel 24 dedicated by the vice-king of Kush, Setau. Photo: © Carola Vogel.

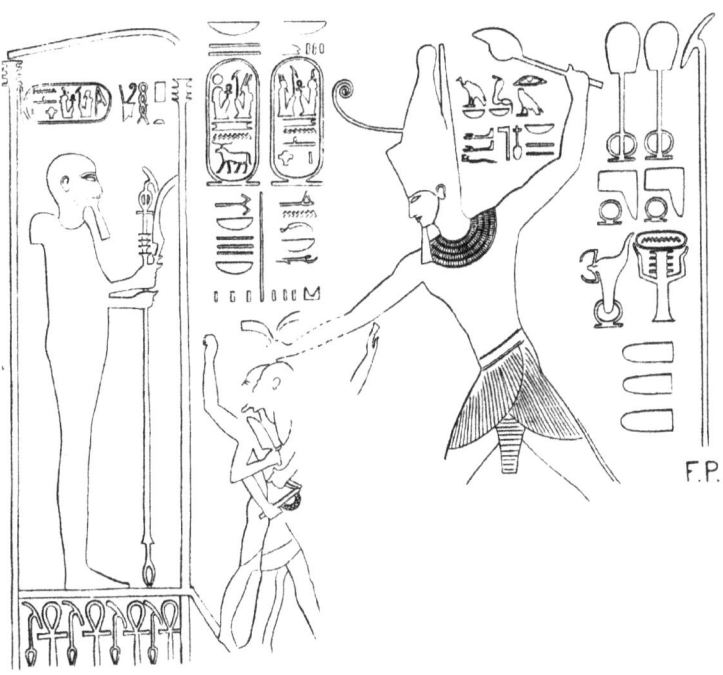

Figure 10: Merneptah smites enemies with the mace-axe before the god Ptah. Lintel from Kom el-Qal'a, right part. After Petrie (1909).

WARFARE AND SOCIETY IN THE ANCIENT EASTERN MEDITERRANEAN

Figure 11: Ramesses III receives the order to go to war. Medinet Habu, 1st Libyan war. After Heinz (2001, 300).

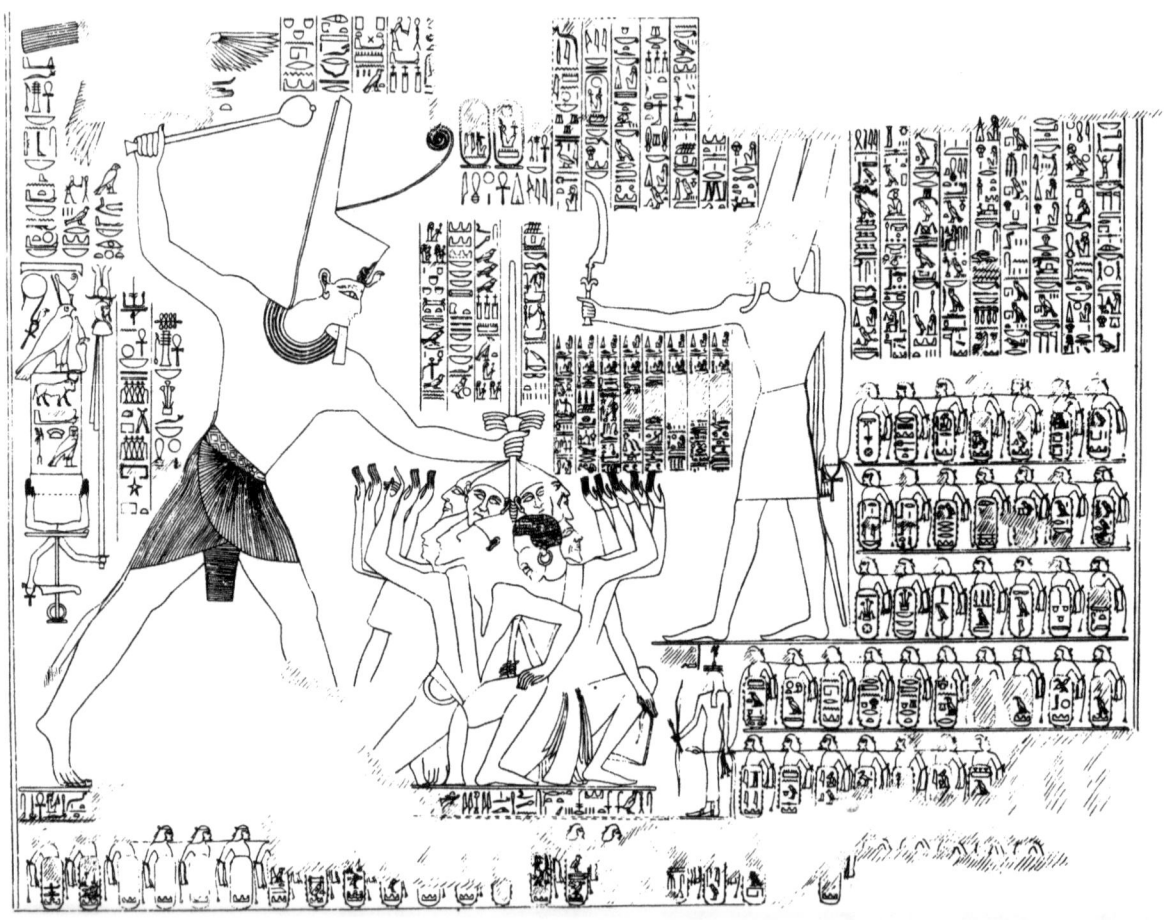

Figure 12: Seti I smiting the nine bows with a mace in front of Amun-Ra, who presents the sickle sword to him. Karnak, northern outer wall of the great hypostyle hall. After Swan Hall (1986, fig. 45).

The text distinguishes the fact that Seti I smote the enemies with his mace but received the sickle sword as a reward.

A second example is attested under Ramesses III. A scene on the second pylon at Medinet Habu (South tower, east wall) is dedicated to the king's war against the Sea People). The king leads three files of tied up captives to Amun-Ra who expects him with a sickle sword (Figure 13) (Heinz 2001, 309).

1.3 The King wears the weapon in selected war scenes

While other groups are worthy of discussion as they give us an idea of the level of significance reached by the sickle sword as an icon of propaganda during the New Kingdom, lack of space demands that I briefly mention only one more here.

The King wears and uses the weapon in selected war scenes. An example comes from Beit el Wali, where Ramesses II threatens a Libyan chieftain with the sickle sword (Figure 14) (Heinz 2001, 259).

1.4 Resume

Thus far, I have concentrated on the fictitious use of the sickle sword in royal context. It has to be considered as one of the most powerful icons of propaganda attested in the New Kingdom. Two main purposes can be extracted:

- The sickle sword as an icon of victory-power transferred from the god to the king.
- The sickle sword as an icon of the pharaoh's unlimited physical strength who smites his enemies.

Before concentrating on those sources, which show the development of the weapon as a standard piece of equipment for infantry and chariot crews, I should take the opportunity to provide some psychological thoughts associated with ancient warfare.

2 Excursus: The Psychology of Killing

Egyptologists – including myself – are sometimes so attracted by the appearance of an artefact and the craftsmanship of those who made it that they are inclined to forget that it served a particular purpose. With respect to the sickle sword, this purpose was the killing of people in close combat. Against this background I would like to include some psychological thoughts by referring to some ideas of Dave Grossman, the author of a landmark book entitled *On Killing: The psychological cost of learning to kill in war and society* (2009).

Grossman notes that on a battlefield every soldier has to overcome some basic instincts such as 'resistance to killing':

At the heart of psychological processes on the battlefield is the resistance to killing one's own species, a resistance that exists in every healthy member of every species. To truly understand the nature of resistance to killing we must first recognize that most participants in close combat are literally "frightened out of their wits". Once the arrows or bullets start flying, combatants stop thinking with the forebrain <...> and thought processes localize in the midbrain, or mammalian brain, which is the primitive part of the brain that is generally indistinguishable from that of an animal <...> The only thing greater than resistance to killing at close range is the resistance to being killed at close range. Close-range interpersonal aggression is the universal human phobia, which is why the initiation of midbrain processing is so powerful and intense in these situations. Thus, one limitation to killing at long range is that greater distance results in a reduced psychological effect on the enemy. This manifests itself in the constant thwarting of each new generation of air power advocates and other adherents of sterile, long-range, high-tech warfare and a constant need for close combat troops to defeat an enemy. (Grossman 2000).

One might ask how soldiers in ancient Egypt were trained to overcome their natural resistance to killing their own species in a face-to-face-situation.[9] Although modern psycho-theoretical approaches are worth considering in trying to understand acts of ancient Egyptian soldiers in close combat, we should be aware that the influence of a different society and cultural background might change our picture tremendously. Even, if we have good reasons to compare the natural resistance to killing across time and region, the question remains how the 'natural' inhibition level fits in. Regardless, one phenomenon applies for every soldier in history: the idea of the survival of the fittest. The better you are trained the better you feel prepared. With respect to the New Kingdom soldier, well-attested training units showed the willingness of soldiers to fight as well as demonstrate their fitness in public combat displays (McDermott 2004, 167 incl. fig. 118, 119). There are also other aspects such as group dynamic processes, giving 'battle chants' some importance. That the success of an Egyptian soldier was judged by the number of enemies killed also provides a better understanding of his motivation.

3 Material Culture

Late Bronze Age sickle swords are known from various places in the Ancient Near East, and are dispersed between museums worldwide. A distribution map for the period in question shows that the weapon is attested up to the Sudan in the south (Ez-Zuma), to Iran in the East, and to Syria in the North (Figure 15). By far the majority of examples come from Canaan.

---

[9] For a discussion of the political legitimization of killing in ancient Egypt, see Assmann (1995).

Figure 13: War against the sea people: Ramesses III leads three files of tied up captives to Amun-Ra, who presents him with the sickle sword. Medinet Habu, south tower, east wall). After Heinz (2001, 309).

Figure 14: Beit el Wali, Ramesses II threatens a Libyan chieftain with the sickle sword. Photo: © Andreas Vogel.

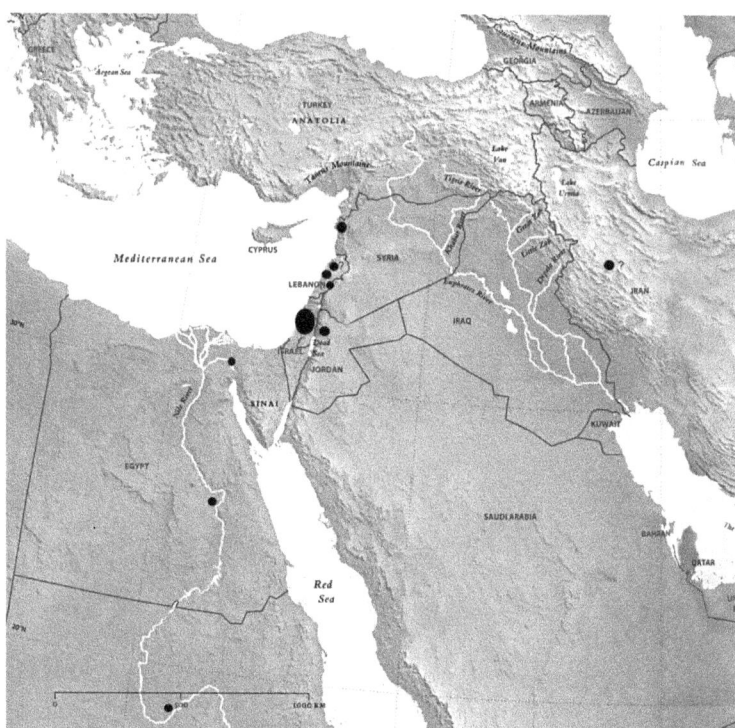

Figure 15: Distribution of the Late Bronze Age Sickle Sword. Map: © Römisch Germanisches Zentralmuseum, Mainz.

The limited distribution of sickle swords observed in the Ancient Near East indicates that they were valuable and will have been melted down for reuse whenever damaged rather than simply discarded. Thus, the few items which are found in tombs refer to the high status of their former owners as members of the elite, who considered it appropriate as a grave good.[10] Sickle swords are known from the Middle Kingdom onwards. Whereas the earlier examples are made of a blade and a separate handle, later items developed into one piece cast swords with inlayed handles.

A weapon of the early type has been discovered at Tell el-Dab'a. It was found in a well-equipped stratum F warrior tomb dating to transitional period MB IIA to MB IIB (Forstner-Müller 1999; 2007/2008). The well preserved blade of the sword is made of copper whereas its handle is made of bone. The blade is cast with a riveted socket, its point voluted. This weapon is the oldest example of this type yet found in Egypt and has a close parallel in an item of unknown origin which was purchased by the Egyptian Museum in Munich (Forstner-Müller 2007/2008, Fig. 4). Both objects recall the Susa-type, but they lack the second volute at the lower end of the blade.

Taking into account the recent find from Tell el-Dab'a, Philip (1989) put the examples from the royal graves at Byblos into perspective. He refers to the ample representational evidence which shows that the voluted type was associated with gods and elites by the later 3rd millennium BC in West Asia. Against this background, the presence of such a weapon at a site where the grave metalwork was very much in a West Asian tradition appears to confirm their non-Egyptian origin. As the Tell el-Dab'a example lacks the Egyptianizing decorative features which were so characteristic of those from Byblos, he considers the latter as exceptional. Thus, he interprets the curved swords from Byblos as local products.

With the early examples in mind we should now proceed to the Late Bronze Age material showing the sickle sword in its highly developed stage.

Bronze swords in general rarely exceeded 800g, and if one weighs over 1kg it becomes too heavy. With respect to curved swords we are sadly lacking sufficient data. The two weights I am aware of are 754g in the case of the Liverpool sword and 800g for the new example recently sold by a Munich art dealer.[11]

The length of the sickle sword ranges from 36cm in the case of the unusual small item from Kamid el-Loz up to 65cm regarding the item sold by a Munich art dealer.[12] All weapons are cast in one piece with an inlayed handle.

---

[10] We know from a scene at the temple of Medinet Habu that sickle swords belonged to that category of weapons of war which had to be kept in royal arsenals and were issued to the army under official control only (McDermott 2004, 121). This fact leads to the question of which kind of weapons the Egyptian soldier was allowed to possess for private use.

[11] I would like to thank Sonia Focke, MA, Munich who drew my attention to this weapon and kindly provided me with its measurements and a photograph. So far this sickle sword has only been published by means of a short notice in the auctioneers' catalogue: Gorny und Mosch Gießener Münzsammlung (2007).

[12] See reference 11.

Shalev's investigations of Late Canaan material found that the blade of the curved sword is up to 25-30% harder than the hilt (Shalev 2004, 84-85). For example, the hardness of the sword originating from the Kefar Samir coast reaches a Vickers Hardness/Hv of 186 at the blade, and 159 at the hilt-base. The difference results from a less complex treatment of the hilt in comparison to the blade after casting.[13]

The actual examples of sickle swords are missing one detail which is known from representations (McDermott 2004, 78, incl. Fig. 78): straps/loops to attach the handle to the wrist, as was done by the sword-knot of more recent times.

As noted elsewhere, I disagree with Hans Wolfgang-Müller's typological attempt to distinguish between the Late Bronze sickle sword as a stabbing and a cutting weapon. Müller considered the two different shaped swords found among the multitude of weapons in the tomb of Tutankhamun as 'representatives for different usage'.

However, the cutting edge of the supposed stabbing sword shows clearly that the weapon was intended to smite the enemy. While long swords are usually designed with a sharp double edge allowing to penetrate the enemy's body up to a lethal depth, the sickle swords in question lack this feature. The sword's point is in most cases not designed for a proper penetration.

Several swords were discovered or properly published only after Müller's studies:

- A sickle sword of unknown origin kept in the Liverpool Museum (Figure 16) (Wernick 2004).
- A sickle sword from the Kefar Samir coast near Haifa kept in the Maritime Museum Haifa (Figure 17) (Shalev 2004, 58).
- A sickle sword from Ez-Zuma, Sudan, now in National Museum in Khartoum (Figure 18) (Żurawski 2002; Davies 2004).
- A scimitar found at Terqa (Masetti-Rouault, Rouault 1996).
- A sword which appeared at an auction house in Munich only recently declared as of Ancient Near Eastern origin (Figure 19).[14]
- A ceremonial sickle sword of unknown origin, Deutsches Klingenmuseum, Solingen 1963.W.100 (Figure 20).

Next, I would like to give some principal technical details on Late Bronze Age sickle swords by discussing the remarkable find from the Sudan (Żurawski 2002; Davies 2004). It has a total length of 58.8cm and reaches a thickness of 6.5cm at the back of the grip. Cast in one piece, it consists of a short flanged handle with a sloping rear end. In the handle are the remains of metal tangs for securing a now missing inlay. Other examples indicate that the inlay was usually made of wood. The sword has an almost straight hilt, flanged on both edges. It curves outwards sharply in the third part of the sword, the curved blade with convex cutting edge and flanged inner edge, ending in a point at the distal end. The hilt and the blade are decorated with a medial rib flanked by furrows probably executed in chased work. This example from the Sudan is closely paralleled by a weapon from Tell el-Rotaba in the Wadi Tumilat (Müller 1987, 158-159), which has been tentatively dated to the 19th Dynasty.

Another remarkable sickle sword is on display in the Deutsche Klingenmuseum, Solingen (1963.W.100). Beside a short catalogue entry (Uhlemann 1964) the weapon has not been discussed thoroughly. The herein published photograph and the provided measurements will hopefully increase awareness of the sword (Figure 20).[15] The item, thought to origin from Luristan, was purchased by the 'Freunde des Deutschen Klingenmuseums' in 1963. Its unusual coiled shoulder is unique within the group of sickle swords, whereas its handle shows the common flanged feature.

An overview of the types – which I prefer to call variations – of Late Bronze Age sickle swords is given by (Figure 21).[16] I have highlighted those from the tomb of Tutankhamun which give us a *terminus post quem*. The same is valid for one example kept in the Louvre which bears the name of Ramesses II. Whereas most of the items are datable only by terms of typological comparison the weapon from Gezer offers at least a datable burial context, which favours a date of LB IIA, or the first half of the 14th century BC.

From iconographic and epigraphic sources it is known that although of western Asiatic origin, the *khepesh* was immediately adopted as part of the standard Egyptian military equipment. It appears as a Syrian tribute in the tomb of Menkhheperresoneb (TT 86 - Thutmosis III), in the tomb of Horemheb (TT 78 - Thutmosis IV), and Meryre II (Akhenaton). However, it had been manufactured in Egypt itself from the reign of Thutmosis III at the latest (Morkot 2007, 182). A scene in the tomb of Qenamun (TT 93 - Amenhotep II) which shows how products of the royal workshops are presented to the pharaoh includes sickle swords, and the caption states their number as 360 (Hallmann 2006, 196).

If one discusses an offensive weapon one should not forget to look at armour also. It seems quite clear that scale armour for man and horse developed in response to the new forms of weapon and warfare practiced in the

---

[13] Whereas the blade was cold worked extensively, annealed and cold worked again, the ridge of the hilt-base was cold worked, annealed and shows traces of slight cold work finish only.

[14] The item shows close similarities to those from Terqa, see Masetti-Rouault and Rouault (1996).

[15] My thanks are due to Lutz Hoffmeister, Deutsches Klingemuseum Solingen who kindly provided me with a new photograph of the object and granted me permission to publish it accordingly.

[16] For an in depth discussion of this typological overview and the corresponding references, see Vogel (2006). The sickle swords from Munich, Solingen and Terqa are excluded, being the starting point of ongoing typological research to be published elsewhere.

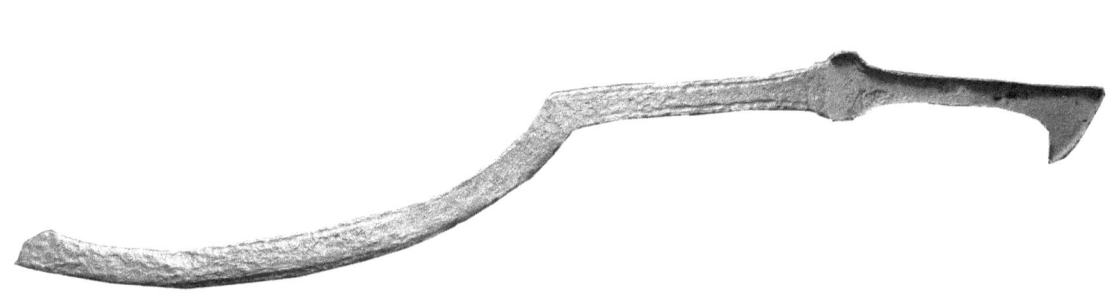

Figure 16: Sickle sword of unknown origin, Liverpool Museum. Photo: Ian Shaw.

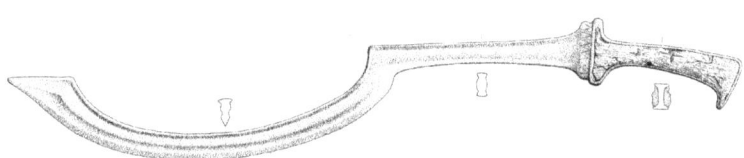

Figure 17: Sickle sword from the Kefar Samir coast near Haifa. After Shalev (2004, pi. 20, 172).

Figure 18: Sickle sword from Ez-Zuma (Sudan). Sketch: Carola Vogel.

Figure 19: Sword of unknown origin, recently purchased by a Munich auctioneer. Sketch: Carola Vogel.

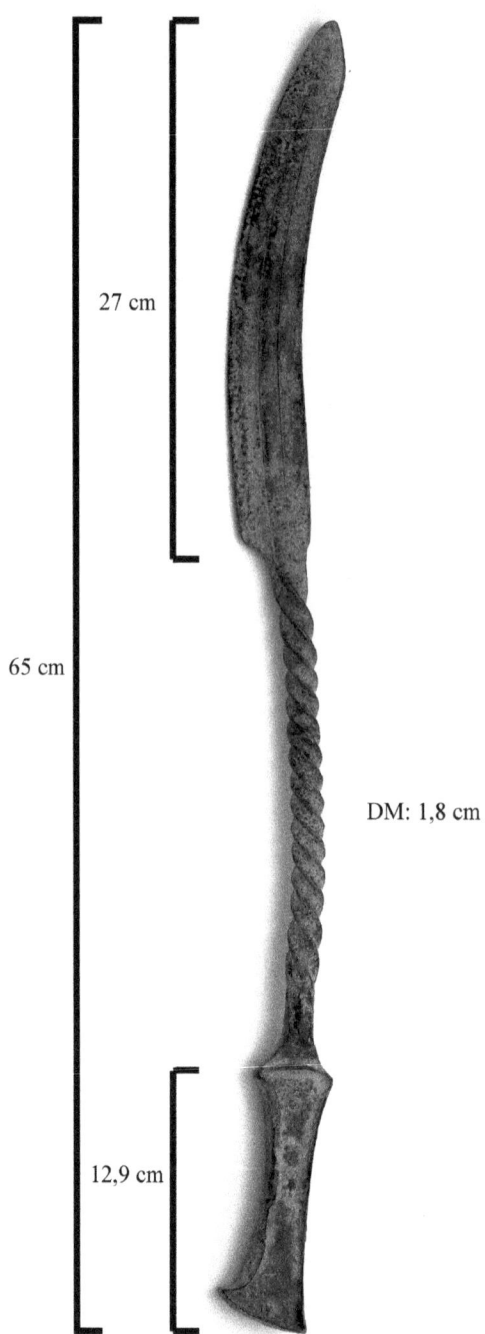

Figure 20: Ceremonial sickle sword with coiled shoulder, Deutsches Klingenmuseum, Solingen 1963.W.I00. Photo: © Deutsches Klingenmuseum, Solingen.

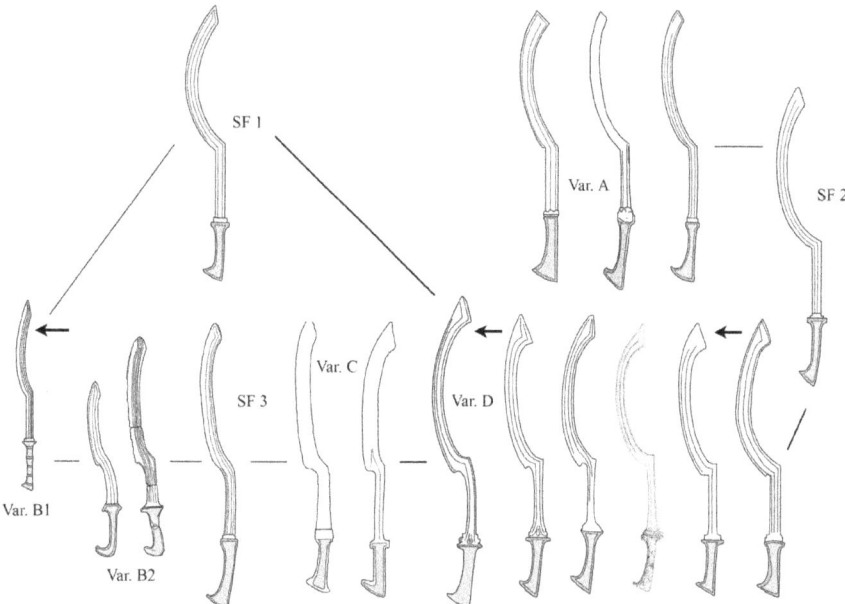

Figure 21: Overview of the types/variations of Late Bronze Age sickle swords. Revision of Vogel (2006, Abb. 7).

Late Bronze Age, such as the sickle sword. With regard to defensive weapons which should protect their bearers against attacks executed with the sickle sword, we have to name three particular items: body armour, helmets, and shields.

Hulit and Richardson (2007, 52-63) have conducted experiments focussing on the construction, manufacture and effectiveness of Late Bronze Age scale body armour from the Middle East. The experiments – which did not include the use of sickle swords but those of archery and axes edged to razor-sharpness – have shown them to be an effective item of equipment. The combined material protected the living targets from harm, thus the use of textile beneath the bronze scales. As known from various illustrations, the use of linen armour seems quite common in the Late Bronze Age.

Whereas bronze helmets appear frequently in New Kingdom representations, there are only rare examples in the material culture record. This can be easily explained by the fact that the bronze objects were extremely expensive and would have been melted down for reuse if no longer needed.

It should not come as a surprise that shields were thought to protect their bearers against attacks executed with the sickle sword. A find supporting this suggestion comes from Qantir. It belongs to a group of engraved lime stones thought to be models for shield fittings. One of them shows a unique motif on both of its sides: a combination of two sickle swords with lotus flowers (Petschel and van Falck 2004, 245-246, Q5 = FZN 1983/0934).[17] The motif is shaped like a loaf of bread and is 27.5cm long and 20.5cm wide. The total depth of the stone is 7cm.[18] The item had been found in the so-called 'multi-functional workshops'. I agree with the interpretation of the motif as a part of a shield fitting, eventually referring to a special unit of its proposed owners. If this is correct, it has to be seen as a small icon of propaganda indicating: 'try to crash my shield it will withstand'.

Acknowledgements

My thanks are due to: Dr Verena-Renate Bach-Berkhahn, Mainz, Dr Vivian Davies, London, Sonia Focke MA, Munich, Dr Irene Forstner-Müller, Cairo, Dr Alfred Grimm, Munich, Dr Dan'el Kahn, Haifa, Dr Volker Grünewald, Mainz, Dr Katja Lehmann, Phoenix, Dr Osnat Misch-Brandl, Jerusalem, Dr Yossi Mizrachi, Haifa, Dr Marcus Müller, Potsdam, Dr Silvia Prell, Cairo, Dr Edgar Pusch, Hildesheim, Dr Sariel Shalev, Haifa, Dr Ian Shaw, Liverpool, Dr Sylvia Schoske, Munich, Dr Deborah Sweeney, Tel Aviv, Dr Andreas Vogel, Worfelden, Dr Irit Ziffer, Tel Aviv.

Bibliography

Assmann, J. 1995. Ägypten und die Legitimierung des Tötens: Ideologische Grundlagen politischer Gewalt im Alten Ägypten. In H. von Stietencron and J. Rüpke (eds.), *Töten im Krieg*. Veröffentlichungen des Instituts für historische Anthropologie E.V. 6, 57-85. Freiburg, Alber.

---

[17] For information regarding its finding situation, I refer to Edgar Pusch, Personal communication 16.03.08: 'FZN 1983/0934, STIN0120. Planquadrat: Q I-f/3, P1.02; 0,78-0,99 NS, 4,42-4,65 WO, 4,38 muNN (UK) - Befund: in tonigem Lehm auf Fußboden zu Mauer M01, unmittelbar über Mauer M01a (gekappt) - Stratigraphie: primar, Stratum B/2a - Einheit: Multifunktionale Werkstatten Raum R05 (bezogen auf das Planquadrat)'.

[18] Edgar Pusch assumes that the rounded and flat sides of the stone were used one after the other, first the rounded one which he explains as resulting from heavy use, whereas the well-preserved flat side would ultimately be used to replace the older side.

Davies, W. V. 2004. Scimitar. In D. A. Welsby and J. R. Anderson (eds.), *Sudan. Ancient Treasures. An Exhibition of Recent Discoveries from the Sudan National Museum*, London, British Museum Press.

Forstner-Müller, I. 1999. Recent Find of a Warrior Tomb with a Servant Burial in Area A/11 at Tell el- Dab'a in the Eastern Nile Delta. *Zeitschrift für klassische Archäologie* 12/IX/99.

Forstner-Müller, I. 2007/2008. A new *scimitar* from Tell el-Dab'a, *Archaeology and History in Lebanon* 26-27, 207-211.

Galán, J. M. 1995. *Victory and Border. Terminology related to Egyptian Imperialism in the XVIIIth Dynasty*. Hildesheimer Ägyptologische Beiträge 40. Hildesheim, Gerstenberg.

Gorny und Mosch Gießener Münzsammlung. 2007. *Auktionskatalog 163 'Kunst und Antike', 14. Dezember 2007*, 222-223. München.

Grossman, D. 2000. Evolution of Weaponry, <http://www.killology.com/print/print_weaponry.htm>

Grossman, D. 2009 (© 1995). *On Killing. The psychological cost of learning to kill in war and society*. New York, Boston, London, Back Bay Books.

Gundlach, R. 2009. Ägyptische Militärgeschichte im Rahmen des pharaonischen Staates: der ägyptische König als „roi de guerre" und „roi connétable". In R. Gundlach and C. Vogel (eds.), *Militärgeschichte des pharaonischen Ägypten. Altägypten und seine Nachbarkulturen im Spiegel der aktuellen Forschung*. Krieg in der Geschichte 34, 49-66. Paderborn, Schöningh.

Hallmann, S. 2006. *Die Tributszenen des Neuen Reiches*. Ägypten und Altes Testament 66. Wiesbaden, Harrassowitz.

Heinz, S. C. 2001. *Die Feldzugsdarstellungen des Neuen Reiches. Eine Bildanalyse*. Denkschriften der Gesamtakademie XVIII. Wien, Verlag der Österreichischen Akademie der Wissenschaften.

Hornung, E. and Staehlin, E. 2006. *Neue Studien zum Sedfest*. Aegyptiaca Helvetica 20. Basel, Schwabe.

Hulit, H. and Richardson, T. 2007. The Warriors of Pharaoh: Experiments with New Kingdom scale armour, archery and chariots. In B. Molloy (ed.), *The Cutting Edge: Studies in ancient and medieval combat*, 52-63. Stroud, Tempus.

Keel, O. 1974. *Wirkmächtige Siegeszeichen im Alten Testament. Ikonographische Studien zu Jos 8,18-26; Ex 17,8-13; 2 Kön 13, 14-19 und 1 Kön 22,11*. Orbis Biblicus et Orientalis 5. Freiburg, (Schweiz), Universitätsverlag.

Lenk-Chevitch, P. 1941. Note Concerning the Distribution of the Sickle-Sword, *Man* 60, 81-84.

Loeben, C. 2008. Neuerungen in Architektur und Relief. In C. Tietze (ed.), *Amarna: Lebensräume, Lebensbilder, Weltbilder*, 254-265. Potsdam, Arcus.

Martínez Babón, J. 1995. *Historia de la espada curva durante el Imperio Nuevo egipcio*. Barcelona, Universitat Autònoma de Barcelona.

Masetti-Rouault, M.G. and O. Rouault. 1996. Une harpé à Terqa. In M. Gasche and B. Hrouda (eds.), *Collectanea Orientalia: Histoire, arts de l'espace et industire de la terre. Etudes offertes en hommages à Agnès Spycket*, 181-198. Neuchâtel, Recherches et publications.

Maxwell-Hyslop, K. R. 2002. Curved Sickle-Swords and Scimitars. In L. al-Gailani Werr, C. Curtis, H. Martin, A. McMahon, J. Oates and J. Reade (eds.), *Of Pots and Plans: Papers on the archaeology and history of Mesopotamia and Syria presented to David Gates in honour of his 75th birthday*, 210-217. London: Nabu.

Morenz, L. D. 1998. Fremde als potentielle Feinde. Die prophylaktische Szene der Erschlagung der Fremden in Altägypten. In H. Preissler and H. Stein (eds.), *Annäherung an das Fremde. XXVI. Deutscher Orientalistentag vom 25. bis 29.9.1995 in Leipzig*. Zeitschrift der Deutschen Morgenländischen Gesellschaft Supplement 11, 93-103. Stuttgart, Franz Steiner.

Morkot, R. G. 2007. War and the Economy: The international 'arms trade' in the Late Bronze age and after. In T. Schneider and K. Szpakowska (eds.), *Egyptian Stories: A British egyptological tribute to Alan B. Lloyd on the occasion of his retirement*. Alter Orient und Altes Testament 347, 169-195. Münster, Ugarit Verlag.

Müller, H. W. 1987. *Der Waffenfund von Balata-Sichem und Die Sichelschwerter*. München, Bayerische Akademie der Wissenschaften.

Müller, H. W. 1990. Zwei weitere Sichelschwerter. In A. Eggebrecht and B. Schmitz (eds.), *Festschrift Jürgen von Beckerath zum 70. Geburtstag am 19. Februar 1990*. Hildesheimer Ägyptologische Beiträge 30, 215-222. Hildesheim, Gerstenberg.

Müller, M. 1998. *Der König als Feldherr. Schlachtenreliefs, Kriegsberichte und Kriegsführung im Mittleren und Neuen Reich*. Tübingen: Unpublished PhD Dissertation.

Müller-Wollermann, R. 2009. Symbolische Gewalt im Alten Ägypten. In M. Zimmermann (ed.), *Extreme Formen von Gewalt in Bild und Text des Altertums*. Münchner Studien zur Alten Welt 5, 47-64. München, Herbert Utz Verlag.

Nelson, H. 1932. *Medinet Habu II: Later Historical Records of Ramses III*. University of Chicago Oriental Institute Publications 9. Chicago: University of Chicago Press.

Petrie, W. M. F. 1909. *The Palace of Apries, Memphis II*. London, School of Archaeology in Egypt.

Petschel, S. and von Falck, M. 2004. *Pharao siegt immer: Krieg und Frieden im Alten Ägypten*. Bönen, Kettler.

Philip, G. 1989. *Metal Weapons of the Early and Middle Bronze Ages in Syria-Palestine*. British Archaeological Reports International Series 526. Oxford, British Archaeological Reports.

Raedler, C. 2003. Zur Repräsentation und Verwirklichung pharaonischer Macht in Nubien: Der Vizekönig Setau. In R. Gundlach and U. Rößler-Köhler (eds.), *Das Konigtum der Ramessidenzeit. Voraussetzungen — Verwirklichung – Vermächtnis: Akten des 3. Symposions zur ägyptischen Konigsideologie in Bonn 7-9.6.2001*, Ägypten und Altes Testament 36,3, 129-173. Wiesbaden, Harrassowitz.

Roth, S. 2002. *Gebieterin aller Länder: die Rolle der königlichen Frauen in der fiktiven und realen Außenpolitik des ägyptischen Neuen Reiches*. Orbis Biblicus et Orientalis 185. Freiburg (Schweiz), Universitätsverlag.

Schoske, S. 1995. *Das Erschlagen der Feinde: Ikonographie und Stilistik der Feindvernichtung im alten Ägypten*. Ann Arbor MI, UMI.

Schulman, A. 1988. *Ceremonial Execution and Public Rewards. Some historical scenes on New Kingdom Private Stelae*. Orbis Biblicus et Orientalis 75. Freiburg (Schweiz), Universitätsverlag.

Shalev S. 2004. *Swords and Daggers in Late Bronze Age Canaan*. Prahistorische Bronzefunde IV,13. Stuttgart, Franz Steiner Verlag.

Sherratt, S. 2000. Circulation of metals at the end of the Bronze Age in the Eastern Mediterranean. In C. F. E. Pare (ed.), *Metals make the world go round: the supply and circulation of metals in Bronze Age Europe: Proceedings of a Conference held at the University of Birmingham in June 1997*, 82-98. Oxford, Oxbow.

Swan Hall, E. 1986. *The Pharaoh Smites His Enemies: A comparative study*. Münchner Ägyptologische Studien 44. München, Deutscher Kunstverlag.

Tawfik, S. 1975. Aton Studies: back again to Nefer-nefru-Aton. *Mitteilungen des Deutschen Archäologischen Instituts Abteilung Kairo* 31(1), 159-168.

Uhlemann, H. R. 1964. *Schwerter und Dolche, Eß- und Schneidgerät der bronzezeitlichen Kulturen*. Solingen, Deutsches Klingenmuseum Solingen.

Vogel, C. 2006. Hieb- und stichfest? Überlegungen zur Typo logic des Sichelschwertes im Neuen Reich. In D. Bröckelmann and A. Klug (eds.), *In Pharaos Staat: Festschrift für Rolf Gundlach zum 75. Geburtstag*, 271-286. Wiesbaden: Harrassowitz.

Vomberg, P. 2004. *Das Erscheinungsfenster innerhalb der amarnazeitlichen Palastarchitektur: Herkunft, Entwicklung, Fortleben*. Philippika 4. Wiesbaden: Harrassowitz.

von Stietencron, H. and Rüpke, J. (eds.). 1995. *Töten im Krieg*. Veröffentlichungen des Instituts für historische Anthropologie E.V. 6. Freiburg, München.

Wernick, N. E. 2004. A Khepesh Sword in the University of Liverpool Museum, *Journal of the Society for the Study of Egyptian Antiquities* 31, 151-155.

Żurawski, B. 2002. Survey and Excavations between Old Dongola and ez-Zuma. *Sudan & Nubia* 6, 73-85.

Post-Traumatic Stress Disorder (PTSD) in Ancient Greece: A Methodological Review

Alan M. Greaves

Abstract

This paper examines the reasons, both medical and historical, why it is impossible to make a conclusive retrospective diagnosis of Post-Traumatic Stress Disorder in any historical character from ancient Greek literature. Medically, these reasons include the changing definitions and diagnostic criteria applied to the condition by the medical profession, and the difficulty of making differential diagnoses between PTSD and other medical conditions. Historically, it includes the potential trans-cultural and trans-historical expressions of PTSD and inherent limitations of our source material. In conclusion, it is proposed here that although conditions akin to what we might today call 'PTSD' were almost certainly common in the ancient world, we should look for evidence of them in the invocation of common literary motifs and tropes rather than by the spurious application of the medicalised diagnostic criteria used by modern psychiatry.

Introduction

Post-Traumatic Stress Disorder (PTSD) is one of the most-discussed psychiatric conditions in relation to ancient literature yet the secure identification of historical figures that experienced it has so far proved impossible. The reasons for this failure to identify it in the historical record of ancient Greece are explored in this paper, which aims not to provide a survey of the possible incidences and references to PTSD in ancient Greek literature, but rather to critically review the reasons why such a survey would be largely in vain. Instead, this paper advocates using a more general understanding of various post-traumatic stress conditions, of which PTSD is just one, and recognizing their prevalence within human populations of all kinds and all periods is a more realistic and helpful way to understand some of the generalizing tropes and motifs of what might be described as 'Old Soldier' characters in ancient Greek literature. This approach will have wider significance for our readings of many different ancient works, be they mythical, literary or historical.

Like the work of Jonathan Shay (1994), this paper is not just a work of academic documentary research, it is also informed by my practice as a psychotherapist. As an ancient historian and archaeologist, I have been interested in the cultural context of warfare in the ancient world (e.g. Greaves 2010); as a psychotherapist I have often worked with clients who experience psychological distress following traumatic episodes in their lives.[1] Both these experiences have come together in the production of this paper. When working with clients who experience post-traumatic stress, it has been my experience that it is best to approach matters holistically, looking at the 'big picture' in order to allow clients to address all aspects of their life prior to, during and after the traumatising event. This therapeutic approach has informed my approach to the subject of traumatic stress in ancient Greece by allowing me to step back from the minutiae of literary and historical criticism and close textual analysis, in favour of a broader overview of the issues concerned.

This is not the first study of PTSD in ancient Greece by any means. In particular there have been important extended studies of just this topic by Jonathan Shay (1994) and Lawrence Tritle (2000) both of which made major contributions to our understandings of both ancient literature and PTSD itself. Other valuable works have also looked more broadly at the role, and effects, of violence in Greek culture and society (e.g. Hanson 1991, 2001; van Wees 2000).

Perhaps the closest any scholar has come to doing this is Tritle (2004), who systematically applied the diagnostic criteria of PTSD to the character of the Spartan general Clearchus in Xenophon's *Anabasis*. He wrote: 'Such a modern interpretation might seem forced, but I believe it is possible to argue with little doubt that Xenophon in fact provides us with the first known case of Post-Traumatic Stress Disorder, or PTSD, in the Western literary tradition' (Tritle 2004, 326). Yet, even given the high level of scholarly interest in the social and psychological consequences of ancient warfare, it may seem surprising that scholars other than Tritle have generally shied away from making direct statements about whether PTSD did, or did not, exist in ancient Greece (e.g. van Wees 2004: 151).

In this paper I will examine why this reticence to draw conclusions about specific incidences of PTSD in historical figures from Greek history is understandable because such an identification is not possible for two fundamental reasons. Firstly, the nature of our Greek historical and literary writings does not support such a direct analysis. The characters we are dealing with, whether mythological or historical, are products of literary genres and the retrospective diagnosis of complex psychological disorders in their descriptions is inappropriate. Secondly, the nature of PTSD as a

---

[1] I began my therapy practice in 2005 and am a registered therapist with the United Kingdom Council of Psychotherpy (UKCP). In my practice I have worked with clients who experience post-traumatic stress or have developed phobias in response to traumatic experiences.

psychological condition in itself means that it would be virtually impossible to identify in any historical context, even if better textual evidence were available. For example, it is difficult to demonstrate PTSD's key diagnostic criteria (e.g. 'flashbacks') in Greek texts, its other symptoms can resemble those of other psychological disorders and the nature of its trans-cultural and trans-historical expression cannot easily be predicted.

It will concluded here that the experience of warfare was so prevalent and so traumatic in ancient Greece that, even if it cannot be formally classed as PTSD using modern medical criteria and terminology, it and other post-traumatic psychological conditions were almost certainly widespread and awareness of these can be seen to have filtered down into ancient works of literature in the invocation of certain literary tropes and motifs. Just as the meaning of 'Shell Shock' has been re-negotiated by contemporary society to inform new understandings of the trench warfare of World War I, so too can it be argued that in ancient Greek literature, the invocation of certain tropes and motifs that were informed by the audience's experience of seeing those around them who had been traumatised by battle came to stand for the psychological damage caused by war. This common understanding, which is implicit in many works of ancient history and literature, can be read to suggest that post-traumatic psychological conditions, far from being uncommon, were actually very common, constituting, and remaining, a universal consequence of war throughout time and this awareness should inform our readings of ancient Greek literature.

A Brief History of the Study of PTSD

'Post-Traumatic Stress Disorder' (or PTSD) is now the accepted term to describe a widely-recognised psychological disorder, but it is known that the condition existed before this current terminology came into use. The observation of a condition that may be identified with PTSD in Western medical literature dates as far back as 1866 (Lamprecht and Sack 2002; citing Erichsen 1866). Understanding of the condition advanced greatly in World War I when 'Shell Shock' became such a problem for the British army that a government paper on the subject was commissioned (Richards 2004; Shephard 2000). Doctors working with those returning from the trenches began to appreciate that because there was no physical cause for the symptoms they were witnessing that could be found to explain Shell Shock, what they were seeing was probably a form of psychological illness that could be treated by means of the new psychotherapeutic approaches then being pioneered by Sigmund Freud and his contemporaries.

At the start of the 20th century, following the publication of his monumental work *The Interpretation of Dreams* (1900), Freud's psychodynamic model of the mind became increasingly influential. Freud continued to adapt his theories for the causation of psychological illness throughout his working life, so as to accommodate into it such events as World War I and the rise of Nazi ideology (Gay 1988, 395-396, 589-596). In *Beyond the Pleasure Principle* (1920), Freud discussed why people that he then called 'neurotics', including those who experienced Shell Shock, repeated unpleasant experiences. As a consequence, he introduced new elements into his psychodynamic model of the human mind in order to accommodate the experiences of victims of Shell Shock, including new introductions such as *Thanatos* (the Death Instinct) and its counterpart, *Eros* (the Life Instinct). Similarly, he also reappraised his theory of the mind to introduce a new element: the Super-Ego (in 1923) to account for the power of social pressure that he had witnessed during that era. Since then, theories about the causation of psychological illness, which for Freud arose from innate psychological structures within the individual, have moved on but Freud's engagement with Shell Shock marks an important point in the development of thinking about the psychological trauma experienced by soldiers during wartime. It gave recognition to the fact that it had a psychological cause and could, therefore, have a psychological cure.

During World War II psychological trauma was referred to as 'Combat Fatigue' (Saul 1945) and in the Vietnam War it was initially known as 'Post-Vietnam Syndrome' (Freidman, 1981; Shephard 2000). In 1980, the terminology of the condition changed again when it was defined as 'Post Traumatic Stress Disorder' by the American Psychiatric Association in the third edition of its highly influential publication the *Diagnostic and Statistical Manual of Mental Disorder* (DSM III). This provided a clearly prescribed list of diagnostic symptoms and criteria for the condition. Its inclusion in the DSM medicalised PTSD for the first time and since then definitions of it have varied, but approaches to its diagnosis and treatment have remained heavily medicalised, probably for medico-legal reasons (such as claims for compensation by those affected).

So far, the Vietnam War remains the most widely studied conflict in relation to PTSD with over 500 papers having been written about it (Kleber, Brom and Defares 1992). Since then, other 20th century conflicts, including Korea and the 1991 Gulf War, have also been the subject of intense interest in relation to the incidence of PTSD, so the predominance of Vietnam in the scholarship of PTSD may soon change.

A wider range of traumatising events have now been identified as being potential causes of PTSD. In addition to military combat, these now include rape, natural disasters, road traffic accidents, torture, and wartime traumas experienced by non-combatant civilians (Holeva et al. 2001; Johnson and Thompson 2008). PTSD can also be caused by social trauma, when the individual feels that their social standing or personal integrity has been violated by feelings of shame, expressions of racism, or catastrophic social embarrassment.

An interesting development in the public perception of PTSD in the later 20th century was its use as a motif in popular music. Public awareness of the terminology and effects of PTSD were now so widespread that they could be adopted by mainstream pop acts in the knowledge that they would be immediately understandable to their audience. For example, there are direct references to 'Post-Traumatic Stress Disorder' in songs such as Paul Hardcastle's *19* (1985) and Sinéad O'Connor's *Famine* (1994). These popular artistic works may provide a model for understanding how the experience of traumatised individuals may have been referenced in works of classical literature in the understanding that the audience of those works would recognise them for what they are (see below).

The psychiatric profession has continued to define and re-define different forms of post-traumatic stress conditions, including PTSD. For example, in 2000 the text revision of the fourth edition of the *Diagnostic and Statistical Manual of Mental Disorder* (DSM IV-TR), five axes of mental disorders were defined, of which PTSD was then said to operate only on Axis 1. When there is evidence of symptoms extending into the other axes (e.g. Axis 2 – long term conditions affecting how the individual relates to the world) then the DSM IV-TR suggests an alternative diagnosis may be appropriate, such as Disorder of Extreme Stress Not Otherwise Specified (DESNOS) (van der Kolk et al. 2005). There is also another version of PTSD, called Complex Post-Traumatic Stress Disorder or C-PTSD, which has yet to be included in the DSM (Roth et al. 1997).

It can therefore be seen that, throughout the history of the 20th century, each generation has re-defined for itself what we may (for now, at least) continue to call 'PTSD' and then to invoke it in its historical writings and art. As Peter Leese has expressed it: 'The memory of Shell Shock is an entirely unstable condition. Like the symptoms of traumatic neurosis, it slips from one part of the collective mind to another, changing its name and its form as surrounding conditions and expectations alter' (2002, 176). The three main motivators for these re-definitions have been the different experiences of, and attitudes to, warfare in each generation, the development of medico-legal procedures that recognise psychological trauma, and the continuing refinement of psychological models, diagnostic tools and treatments by the medical profession. The process by which each generation negotiates the meaning of 'PTSD' for itself is therefore likely to persist into the future, not only in its military, legal and medical procedures, but also in its art and popular culture.

Differential Diagnosis: PTSD and Other Disorders

The shifting definition of PTSD in medical and general usage demonstrates how complicated it was to define the condition even in 20th century Western culture. However, when we start to approach its identification and definition in cultures that are more distanced from our own, both culturally and temporally in the case of ancient Greece, these complications are compounded.

The most secure way to demonstrate that some version of the psychological condition currently known as PTSD existed in antiquity would be to find clear and unequivocal evidence for its symptoms appearing in historical accounts of the time. However, this is easier said than done, as the symptoms that are currently used to diagnose it are of a type that is common to a number of conditions. As Jonathan Shay wrote: 'PTSD can unfortunately mimic virtually any condition in psychiatry' (1994, 168-169).

As it is currently defined, PTSD has a complex set of symptoms, many of which it shares with other disorders. Symptoms can include:

Re-experiencing the trauma in dreams, recurrent thoughts and images (i.e. 'flashbacks').

- Depression.
- Anxiety disorders.
- Somatisation disorder.
- Violence and criminal behaviour.
- Loss of memory and perception.
- Hypervigilance (veterans are in a persistent state of combat readiness).
- Exaggerated startle response (veterans can activate combat skills in civilian life).
- Destruction of social trust.
- Preoccupation with the 'enemy' (veterans can see 'reds under the bed').
- Alcohol and drug abuse.
- Suicidality, despair, isolation.
- Subsequent chronic health problems.

With such a long and specific set of diagnostic criteria there is a danger that we risk reducing individuals' experiences of PTSD to just a checklist of symptoms. A broader, and for the historian perhaps more workable, summary definition is provided by Stirling and Hellewell (1999, 86-7):

- Persistent symptoms of anxiety.
- Avoidance behaviour.
- Phenomena, such as intrusive and vivid recollections or disturbed dreams, that reflect the involuntary re-experiencing of the traumatic event.

Identifying these symptoms in ancient literature is a more complicated matter that it might at first appear. For example, the single most important criterion for a diagnosis of PTSD is without doubt the re-experiencing of the initiating trauma in dreams, recurrent thoughts and images (Reber and Reber 2001, 551). These 'flashbacks' can take many forms but they must involve the repeated re-experiencing of the originating stressor (DSM IV-TR, 463, 468 - Criterion B). The individual must experience these to differentiate it from other conditions, such as Obsessive Compulsive Disorder, Schizophrenia,

Substance-Induced Disorders, or other psychotic, mood or general medical conditions (DSM IV-TR, 467). The onset of PTSD can also occur many months after the precipitating event (Stirling and Hellewell 1999, 85-87). For the historian, proving that such flashbacks occurred and linking them to the initial trauma will be very difficult given the nature and limitations of our ancient sources.

As noted above, the DSM IV-TR defines five axes of mental disorders and, as a clinical disorder including all the symptoms listed above, PTSD only operates only on Axis 1. However, where such symptoms can be identified in Axis 2, that is long term conditions affecting how the individual relates to the world, then the PTSD is having a more 'global' effect on the client's operation and it may have begun to affect deeper elements of their personality. Alterations in Axis 2 might include:

- Regulation of affect and impulses.
- Attention or consciousness.
- Self perceptions.
- Relationships with others.
- Systems of meaning.

For example, if the individual is experiencing flashbacks and nightmares as the result of a traumatic event but it does not affect them as operational individuals, then a standard diagnosis for PTSD would apply (e.g. a car crash followed by nightmares and reluctance to get in a car). If, however, it affects the individual's broader functioning, then a diagnosis of DESNOS may be more appropriate and treatment is necessarily more involved (e.g. if the individual was a victim of violent assault and frequently disassociates themselves and cannot leave the house, go to work, or otherwise function normally). The symptoms of DESNOS and PTSD can clearly overlap and differentiating them in psychotherapy clients can be tricky, but in literary accounts of deceased or fictional individuals, it will be impossible.

Diagnosis of PTSD is made particularly difficult when direct interaction with sufferers is lost because there is a crucial temporal dimension to its diagnosis. For example, if the symptoms develop and resolve themselves within four weeks of the stressor event, then the diagnosis would be one of Acute Stress Disorder, not PTSD (DSM-IV-TR, 469-472). Also 'symptoms of avoidance, numbing and increased arousal that are present before exposure to the stressor do not meet the criteria for the diagnosis of PTSD and require consideration of other diagnoses (e.g. a Mood Disorder or other Anxiety Disorder)' (DSM-IV-TR, 467). Instances where such temporal observations of an individual can be made in ancient literature are rare.

It is also important to understand the precise nature of the 'flashbacks' that are such an important criterion for successful diagnosis, so as to differentiate it from Obsessive Compulsive Disorder, Schizophrenia, Substance-Induced Disorders, or other conditions (DSM-IV-TR, 467). The flashbacks associated with PTSD can take many forms, such as dreams, but they *must* involve the repeated re-experiencing of the originating stressor (DSM-IV-TR, 463, 468 - Criterion B). Such internal thought processes can only be shared by the individual themselves, for example during counselling, and the nature of the ancient literary sources that we have is such that they are unlikely to give us this level of detail about the internal thought process of the individuals being portrayed. Kruger noted that researchers of 'dead' languages cannot be 'participant' observers in the way that anthropologists (or counselors or therapists) can be (2005, 187-188) and the ancient historian Neville Morley wrote: 'psychoanalysis normally relies on hours of conversation with the patient about their memories and feelings, not on second-hand accounts from historians with axes to grand' (2004, 112).

In addition to PTSD, DESNOS and Acute Traumatic Stress (discussed above), there are other psychological conditions that can result from exposure to a traumatic stressor. These include Panic Disorder and Conversion Disorder.

Panic Disorder is common in victims of mass violence and it can also be a feature of PTSD, so differentiating between the two when making a diagnosis can be difficult (Hinton, Pitch and Pollack 2005). In the panic attacks associated with PTSD there will be a trigger – a loud bang, for example – whereas in Panic Disorder there will be a triggering scenario, such as enclosed spaces (Hinton, Pitch and Pollack 2005, 38-39). It is therefore necessary to discuss closely with the client, in counselling, what their individual history, stressors, and triggers are. Such investigation is, of course, impossible when trying to make a retrospective diagnosis based on ancient literature.

Conversion Disorder, which is also known as Somatoform Reactions or Hysterical Conversion Reactions, is a condition in which psychological responses to a stressor are somatised, resulting in a physical manifestation in the body (Weintraub 1983). An example of such a somatic reaction would be paralysis in a limb for which there is no physical cause, but which is the result of a psychological response to traumatic stress. In such cases, there is a 'close temporal relationship between a [psychological] conflict or stressor and the initiation or exacerbation of a symptom' and this may be helpful in this determination (DSM-IV-TR, 494).

In addition to the fact that many of the medical definitions of different post-trauma stress conditions share symptoms with other non-traumatic disorders and with one another, it is also possible that individuals were experiencing more than one condition simultaneously. The possibility of co-morbidity, whereby individuals may have suffered PTSD and another psychological or physical condition that displayed similar symptoms makes it hard, if not impossible, to ever securely argue for cases of PTSD in antiquity.

There are, therefore, a number of different conditions that share symptoms, and differentiating between them is difficult and requires close consultation with the living individual in order to establish causational, temporal and ideational relationships between their symptoms and any originating stressor. This makes the identification of PTSD, by the terms of reference used in contemporary psychiatry impossible when working in historical contexts. However, there is one case from Greek history when a *post-factum* diagnosis of a traumatic stress condition may, at first, appear to be possible and this is the case of Epizelos at the Battle of Marathon.

Epizelos at Marathon

The case of Epizelos son of Kouphagoras is the best documented example of any form of traumatic shock in ancient Greek historical writings. Epizelos was apparently struck blind during the Battle of Marathon in 490 BC, without a blow ever being laid upon him, when a large bearded Persian killed the man immediately next to him (Herodotus, 6.117). As van Wees put it: 'Blind terror is only a figure of speech to most of us, but some soldiers are so traumatised by combat that they are literally struck blind. One such casualty was Epizelos' (2004, 151). This 'blind panic' is entirely consistent with a diagnosis of Conversion Disorder, the symptoms of which can include 'sensory symptoms or deficits [which] include loss of touch or pain sensation, double vision, blindness, deafness and hallucinations' (DSM-IV-TR, 493). As Lawrence Tritle has written: 'What happened to Epizelos ... has been experienced by other soldiers too in the modern era – hysterical blindness, in which the mind intervenes to protect the body from the horror confronting it' (2006, 214-215; also 2000, 64, n. 24).

The causes of Epizelos' blindness have been disputed and misunderstood by academic commentators (Lazenby 1993, 80). For example, Victor Davis Hanson wrote that Epizelus' Persian was an *epiphany* – a divine vision of a phantom brought on, in his case, by battle fatigue (1989, 192-193; Tritle 2000, 63-65). However, in a recent detailed commentary on the episode, Lionel Scott examined all the possible medical alternatives to the 'obvious' explanation of hysterical blindness, but found little conclusive evidence to support any physical medical explanation (2005, 395-396).

In the intensity of battle, Tritle argued, Epizelos' mind was protecting him from the 'ferocity and chaos' of the fighting at Marathon (2006, 214-215). Whereas in World War I survivors of the trenches frequently experienced hysterical deafness, presumably to block out the sound of bombs, Greek *hoplite* soldiers' responses to the horrors they saw with their own eyes during the mêlée of battle was evidently blindness. In addition to the case of Epizelos, we might also consider the cases of two Spartan soldiers who were affected by blindness prior to the battle at Thermoplyae (Herodotus, 7.229.1), but where medical explanations such as conjunctivitis or trachoma can also be ruled out (van Wees 2004, 151, n. 1).

Based on this, there would appear then to be some evidence to say that, during the classical period at least, there are reported instances of blindness of at least one and possibly as many as three Greek *hoplite* soldiers that are consistent with Conversion Disorder associated with traumatic stress. However, it is not such a simple matter to make such a firm assertion about PTSD for two important reasons. Firstly, the symptoms of Conversion Disorder, such as blindness, deafness and paralysis, are physical and therefore objectively observable to others and describable by them in writing. It should also be noted that somatisation is a common feature of both PTSD and Conversion Disorder and so differentiating between these two conditions in our historical sources, even in the well-attested the case of Epizelos, is difficult. In PTSD, the symptoms are psychological and behavioural in nature and most would need to be described to writers by the sufferer, in particular the key diagnostic criterion of flashbacks. Secondly, the long list of diagnostic symptoms of PTSD, such as depression, anxiety and loss of memory, are all common to at least one or more other psychological conditions and some physical ones.

PTSD as Trans-cultural and Trans-historical Phenomenon

Padmal de Silva wrote: 'it is very clear that the vulnerability to PTSD is not culturally limited' (1999, 125). That is to say, it affects people of all cultures and, presumably therefore, of all periods. Lawrence Tritle also concluded: '...I would argue that the reactions of human beings to the effects of violence have changed little from the time of Xenophon and Clearchus to the present' (2004, 329). This is almost certainly true, as *homo sapiens* achieved full behavioural modernity 50,000 years ago and our species' physiological and psychological responses traumatic stress are likely to have remained largely the same since then. However, the precise form of those responses and the presentation of symptoms of PTSD, as well as the other post-traumatic illnesses outlined above, varies considerably according to cultural factors. Such cultural factors can change not only between geographical regions, but also between sub-cultures within the same society and over time.

It is important to recognise this when we start to look for potential evidence of PTSD in ancient literature because we cannot make the blanket assumption that the symptoms that are observed in PTSD in contemporary Western culture would necessarily be those that were expressed in ancient Greece. Indeed, modern community studies have shown that the incidence and expression of PTSD can vary considerably between different contemporary cultures.

These differences may be connected with cultural beliefs surrounding the aetiology of the conditions. For example, in Panic Disorder sufferers of Cambodian origin it is conceived of as a problem with their *Khyâl* (a metaphorical and spiritual 'wind') which is thought of as

a key part of the physiology of the body in that particular culture (Hinton, Pitch and Pollack 2005, 44-47). It would be interesting to know how ancient societies conceived of the causes of such disruptions in relation to their concepts of the flow and balance of the bodily humours and contemporary concepts of the causation of disease. However, such a study is beyond the scope of this paper, and quite probably beyond the scope of the surviving literary evidence.

In the Vietnam War there were noticeable differences in the incidence of PTSD between White, Black and Hispanic combatants in the American forces (de Silva 1999, 119-120, citing Kulka et al. 1991). This may be accounted for by differing cultural factors within and between the different ethnic groups and sub-cultures that exist within United States society.

The display of symptoms that might be taken as diagnostic expressions of PTSD will also be determined by socio-cultural factors, such as social taboos on men crying or the abrogation of suicide (de Silva 1999, 129-130). For example, de Silva wrote: 'Afghans, like many other people from Oriental cultures, tend to somatise emotional problems' (de Silva 1999, 121). That is to say that in Afghan society it is common to give physical expression to psychological states. However, we do not know how common or acceptable somatisation or the outward expression of any of the other psychological symptoms of PTSD would have been in ancient Greece, which in itself was made up of many differing local cultures and communities.

Having considered the incidence and expression of PTSD between cultures, we must also consider how it may have varied across time. We cannot simply assume that it was necessarily a trans-historical phenomenon; nor can we assume that its manifestations in the past would be recognisable to us in the terms by which it has been defined in contemporary Western medical literature.

Fischer and Manstead noted: 'there are both cross-cultural similarities and differences in emotion' (1996, 240). There is, therefore, no universal human reaction that can be predicted in all circumstances of traumatic stress, but neither are our reactions entirely culturally determined. Bearing in mind the very different incidences and presenting symptoms in different cultures, can we possibly ever know what PTSD, or any of the other post-traumatic conditions discussed above, would have looked like in ancient Greece? Indeed, is PTSD simply a product of modern styles of warfare that may never have existed in any way that we might recognise in the ancient world (van Wees 2004, 151)?

However, the fact that PTSD has been identified, to various degrees and variously expressed, across different ethnic and cultural groups does indicate that the human species is naturally predisposed towards reaction formation following episodes of traumatic stress. As discussed above, there would appear to be some evidence that Conversion Disorder was present in Greece in the 1st millennium BC, even if finding similar evidence for PTSD is a more complicated matter because of the nature of our sources and its symptoms and definition. The sheer prevalence of PTSD across modern populations (see below) adds further weight to the argument that it has been a feature of human experience for a considerable period of our history, even if the precise form of its expression has varied across contemporary and historical cultures.

Epidemiology of PTSD

Having considered the challenges of making a differential diagnosis between PTSD and other psychological conditions and the differing cultural and trans-historical expressions of the condition, let us now consider its epidemiology – that is, its occurrence within the population as a whole.

The DSM (discussed above) concludes:

> community-based studies reveal a lifetime prevalence for PTSD of approximately 8% of the adult population in the United States…studies of risk individuals (i.e. groups exposed to specific traumatic incidents) yield variable results, with the highest rates (ranging between one third and one half of those exposed) found among survivors of rape, military combat and captivity, and ethnically or politically motivated internment and genocide. (DSM-IV-TR, 466).

In other studies 15% out of a sample of 1,600 male survivors of the Vietnam conflict were shown to meet the criteria for PTSD (de Silva 1999, 119-120, citing Kulka et al. 1991). A Harvard University project that aims to co-ordinate a global mental health policy for the victims of mass violence has been called 'Project One Billion' because this is the number of people affected by violence today (McDonald, Bhasin and Mollica 2005, 313). PTSD can therefore be seen to be a very common experience for survivors of conflict, violence and disaster – but if we are to consider the possible incidence of PTSD in antiquity, it is necessary to consider the different nature of war in the pre-industrial era.

World War I marked the start of a new form of warfare and on a scale never previously seen. The conditions of trench warfare were particularly horrific and there was also low morale in the trenches, poor medical services, huge death tolls, and poor treatment of the psychologically traumatised – who were often executed as deserters. However, as Hans van Wees put it: 'The trauma of ancient Greek battle was different from the experiences which leave so many modern soldiers 'shell-shocked' or debilitated by PTSD. Greek soldiers rarely came close to suffering the extremes of physical deprivation associated with trench or jungle warfare, and never saw their friends blown to pieces. On the other hand, *hoplites* suffered the devastating experience...of standing at no more than arm's length from the enemy and laying into one man after another, with spear, sword, and ultimately bare hands and teeth' (2004, 151).

Hoplite warfare may not have been without its horrors, but there were many other variables that make comparison to modern epidemiology studies hard and these make it difficult to predict the potential incidence of PTSD across populations in the ancient world. For example, the reality of the violence of *hoplite* warfare would have been affected by the individual's interpretation of it. When working with clients who experience PTSD in contemporary psychotherapy, it is important to assess the degree to which the pre-morbid individual was a functional individual, or not, as this is likely to affect their ability to recover from the trauma (G. Ibbotson personal communication). Factors such as these are unique to the individual concerned and cannot easily be taken into account when looking at the battle experience collectively or retrospectively.

The above quote from van Wees shows that ancient warfare could, in its own way, be as brutal as modern warfare. Kurt Raaflaub has also argued that it was also a much more common experience across the population as a whole than warfare is today. As he recently wrote: '[In Rome]...between 197 and 168 BC on average 47,500 citizens (out of a total population of c.250,000 adult male citizens) fought every year in long wars abroad; if we applied the same ratio to the USA, many millions of Americans would be fighting for their country every year' (Raaflaub 2007, 9). Indeed, it has long been recognized that many ancient societies were in a state of almost 'perpetual war' (Morley 2000, 170, citing Hume). A high proportion of ancient populations were therefore likely to have been exposed to traumatic events that were, in their own way, as violent as those that are known to result in psychological reaction formation in modern populations today.

Like modern wars and disasters, ancient warfare also affected the general population, not just men of fighting age. Research among Bosnian refugees revealed that there were increased rates of PTSD, depression and anxiety in the adult and child population, and co-morbidity of depression with physical disability resulting from the violence (McDonald, Bhasin and Mollica 2005, 306). 70% of Kuwaiti children affected by the Gulf War exhibited some symptoms of post-traumatic stress (McDonald, Bhasin and Mollica 2005, 306) and Cambodian and Vietnamese refugees to the US also experienced a rate of 70% of PTSD and 50-60% Panic Disorder (Hinton, Pitch and Pollack 2005, 38-39). van Wees has also discussed the conditions and treatment of prisoners of war and refugee populations resulting from ancient wars, which is a subject that has often been overlooked by scholarship on ancient warfare because of the nature of our sources, which tend to focus on the elite males fighting in the *phalanx* rather than the effect of war on the population as a whole (van Wees 2004, 148-149).

Social factors play a part in individuals' reactions to their wartime experiences and the likelihood that they will develop PTSD as a result. They are known to have affected the prevalence of Shell Shock during World War I – in particular the involvement of the media, politics and the pressure to enlist (Leese 2002). Such social pressures also existed in the ancient world, as Matthew Christ's studies of conscription and draft-dodging in classical Athens demonstrate (Christ 2001; 2004; 2006). The intensity of the relationships that existed between the men in a citizen army like the classical Greek *phalanx* would also have intensified the trauma of seeing fellow soldiers cut down, as the victims had been the close friends, neighbours, relatives and possibly even lovers of the traumatised individual left behind.

When morale within military cultures is high, incidence of combat stress reactions are reduced and vice versa when morale is low (de Silva 1999, 128 citing Labuc 1991, 485). The perception of social support (or its removal/destruction) for the trauma victim will also affect the incidence of PTSD. As de Silva wrote: 'This support can contribute to the reduction of the probability of the individual developing full-blown PTSD, and also to the speed of recovery and adjustment' (de Silva 1999, 127). In Perikles' funeral oration he makes it clear that Athenian citizens were born and raised to fight (Thucydides, 2.34-46). *Prima face*, this might appear to suggest that there was a generally supportive attitude to the role of soldiers in society, but Christ's work suggests that there were incidents of individual dissension from this and the Periklean view was not universally held.

Discussion

In the 20th century there was a move to medicalise the psychological illnesses of war veterans, starting with Shell Shock, and particularly surrounding the experiences of veterans of the Vietnam War. This medicalisation may, in part, have been driven by a medico-legal culture that centred on compensation claims and issues of political accountability. Ethnographic studies with non-Western populations, and sub-cultural groups within Western cultures, show that PTSD is a trans-cultural phenomenon but that it manifests itself differently and at different rates between societies and individuals according their cultural and personal frames of reference. This being so, we must seek to avoid reductionist approaches to the study of PTSD that reify human experiences of mental illness and war to simplistic checklists of criteria, against which we can read classical literature in the hope of making a retrospective pseudo-diagnosis based on a set of symptoms that are specific to the modern Western experience. This would be an inappropriate methodology for two reasons.

Firstly, it misrepresents the modern psychiatric process of diagnosis. As outlined above, there are multiple conditions that can result from exposure to traumatic battle conditions – PTSD, DESNOS, Acute Traumatic Stress, Hysterical Conversion Reaction, Panic Disorder, etc. – and to define these in an individual requires more than just diagnosis against a checklist of symptoms, but also observation over time and meaningful interaction with the individual and their inner thought processes – none of which are possible when done remotely and retrospectively.

Secondly, it misrepresents the historical processes at work. Writers of ancient historical and literary works, such as Homer, Herodotus, and Xenophon, were consciously composing works of literature; they were not producing accurate blow-by-blow accounts of historical events as they happened. Rather, they were constructing over-arching narratives within which the specific events and personalities that their audience might be interested in were depicted in a way that fit into their larger story. This is as true of contemporary writers of ancient history as is of the ancient authors that they themselves cite (Morley 1999). Even writers of biographies, such as Plutarch, can be seen to have been consciously weaving biographical narratives that incorporated contemporary philosophical concepts (de Blois 2005).

Given that our source materials, whether consciously 'artistic' or supposedly 'historic' in nature, were the products of some form of literary tradition it may be more appropriate to find different ways of reading them to find indications of the existence of PTSD rather than by the spurious application of a checklist of symptoms looking for the medical 'facts' against which to make a diagnosis. One literary technique that appears to have been used to depict mental illness in other ancient literature is that of using a single motif that is then taken to represent a more complex set of symptoms, emotions or behaviours. For example, in Near Eastern literature a single *topos* of 'aimless, repetitive locomotion' or 'wandering about' is used to evoke depressed mental states, of which such psycho-motor agitation is a just one of many possible symptoms (Barré 2001; Kruger 2005).

In relation to classical literature, Ruth Padel has demonstrated that 'wandering' was also a literary or linguistic motif used in the depiction of madness throughout classical and medieval cultures (1995, 107-119). When applying this same idea of a single 'diagnostic' literary motif that might be taken as emblematic of the full suite of complex symptoms that have come to be associated with PTSD (see above) no obvious single symptom presents itself in the literature. Although 'flashbacks' are a unique symptom of PTSD, they are unlikely to appear in ancient literature, for the reasons discussed above.

However, could be argued that an excessive love of battle might be used to stand as such a *topos* or literary motif. Although 'love of battle' *per se* is not in itself a criterion for a diagnosis of PTSD, as a literary motif it can be seen to describe a number of the recognized symptoms of the condition, including violent behaviour, hypervigilance, exaggerated startle responses and preoccupation with the 'enemy' (see above). I now will examine two possible examples of this motif at work in two very different genres of Greek literature – Xenophon's portrayal of Clearchus in his historical work *Anabasis* and Aristophanes' depiction of Lamachus in his comic play *Archarnians*.

Xenophon's biographical portrait of Clearchus in *Anabasis* appears to show him displaying a great love of battle. Even in a violent and highly militarised state like classical Sparta (Hornblower 2000), such excessive love of war as was demonstrated by Clearchus might be deemed excessive (Tritle 2004, 326). Given Clearchus' long military career, it is not unreasonable to assume, as Tritle (2004) did, that his supposed 'love' of was the result of extensive exposure to violence during his lifetime and a possible expression of PTSD.

Shifting to Aristophanes' portrayal of the Athenian general Lamachus' in *Acharnians* (first staged in 425 BC), it is hard to show such a clear association because of the conflicting nature of historical reference to Lamachus as a public figure and the poetic structure of the comedy that Aristophanes is constructing around him. When he first appears on stage already fully-armed (at line 572), Lamachus is the very 'personification of war' (MacDowell 1995, 170). He would already have been known to a contemporary Athenian audience as someone who, despite his extensive first-hand experience of battle, was seemingly 'eager for the continuation of war' (Sommerstein 1980, 184). However, that fact that he swore an oath of peace with Sparta in 421 BC would appear to be somewhat at odds with this perception (Thucydides 5.19.2), as would his presumed motivation of making money out of the war (Plutarch *Nicias* 15.1, *Alcibiades* 21.9 *contra.* Aristophanes *Archarnians* 595-7, *Peace* 304, 473-4, 1290-4). Also, it is clear that Aristophanes is playing up Lamachus' lust for war so that when the character reappears on stage he can construct an *antistrophos* (a poetic structure of opposites, at lines 1190-1235) between Lamachus's lamentations for having pursued the course of war, against the celebrations of the play's main character Dikaiopolis, who is simultaneously reaping the benefits of peace on the other side of the stage (Rogers 1910, 181).

Given the nature of their literary context, it is neither possible nor worthwhile speculating as to whether or not Lamachus, or even Clearchus, as historical individuals ever experienced PTSD in reality. However, the portrayal of these two historical characters have one thing in common which is that despite having both been closely and repeatedly involved in violent conflict, they are both still depicted as having a perverse-seeming 'love of war'. The fact that they can be portrayed in this way to these authors' respective audiences may reflect a deeper understanding of the lived experience of generations of soldiers and their personal reactions to traumatic stress, which may result in a seemingly counterintuitive obsession with war.

What is telling in these depictions is not that the behaviour of these 'Old Soldier' characters belies any particular instance of PTSD, but rather that such a trope, as a standard set of meanings and interpretations, drew its power from the fact that the audience of these works could identify with it from their personal experience of seeing those around them who had returned from war, changed. As Tritle put it: 'The psychic toll of battle ... can turn a survivor like Clearchus into a grim, solitary warrior' (2000, 63). The authors of these ancient literary works knew this to be true and therefore wrote it into their works in the knowledge that their audience knew it too. Their personal observations of war veterans of both author and audience were therefore sublimated into the construction of the archetype of the 'Old Soldier'. In a similar vein, I have recently argued that the same may be true of Post-Natal Depression, subtle references to which might belie its true prevalence in ancient Greek society (Greaves 2009). Like Post-Natal Depression, PTSD may have been some a commonplace element of the people's lived experience in ancient Greece that more overt references to its existence and effects were not needed.

Another depiction of an individual traumatised by the experience of war is the Homeric construct of the mythical character of Achilles in the *Iliad*. In creating this poetic work the Homeric composer has evidently mapped his characterisation of the fictional Achilles onto the battle experience of traumatised soldiers and composed a fictional work that follows the stages of an individual's psychological reaction to battle trauma. Crucially, in the *Iliad* we see how the character of Achilles is *changed* by the experience of war – giving us the essential *temporal* dimension that is so often lacking in other ancient works (Shay 1994). Although this means that we cannot 'diagnose' the Homeric Achilles with PTSD, because he was not a real person (unlike Clearchus and Lamachus), it does allow us to recognise that such a portrayal must have struck a chord with the audiences for whom the Homeric works were composed and that PTSD-like traumatic stress conditions were probably such a widely-observable fact of life in ancient Greece, that the realisation of their existence permeated the works of many writers, either consciously or inadvertently.

In a brilliant comparative analysis of the characterisation of Achilles and the experiences of Vietnam veterans experiencing PTSD, Jonathan Shay demonstrated that although it would be inappropriate to attempt to clinically diagnose the character of Achilles with PTSD, the depiction of this character reflected many of the same motifs that Shay saw in the testimonies of his own PTSD patients (Shay 1991, 1994). The themes that Shay explores through this unique approach are not bound by a specific list of diagnostic medical criteria, but rather it explores the lived experiences of Vietnam veterans and relates these to the Homeric character of Achilles. So, for example, themes that are important to Shay's veterans and which can also be seen in the story of Achilles include things such as 'the betrayal of "what's right"', which was evidently highly emotive for them, but which goes beyond reductionist medicalised diagnostic criteria.

But understanding PTSD in this way is not just a means to interpret and rationalise the past and seek to make objective sense of it; it is also a way to negotiate the past's meaning for contemporary society. Peter Leese, for example, has written about: 'The post-1989 revival of interest in Shell Shock as a metonymic symbol of the war' (Leese 2002, 172). Just as Leese proposes that Shell Shock is a 'prism' through which to view World War I (2002, 176), so too might we use the literary trope of the war-loving veteran, or the 'Old Soldier', to read new meanings into ancient literature, not just for their significance in ancient society, but also to negotiate a new meaning for them for our own.

Knowing the contemporary diagnostic criteria for PTSD can indeed extend our understanding of the 'truth' behind the depiction of figures in Greek history and literature, as Shay did for Achilles (1994) and Tritle did for Clearchus (2004), but both Shay and Tritle go beyond simply deepening our literary criticism of ancient works by means of dry textual analysis. Shay used the *Iliad* in his therapeutic work with patients experiencing PTSD, in a form of therapy called 'Milieu Therapy'. Tritle also reported that two Vietnam veterans who attended his history classes were able to find a degree of solace in identifying with the psychologically wounded character of Clearchus. He wrote: 'The Vietnam veterans [in his class] saw Clearchus as in some ways reassuring. They saw that their own anger and bitterness induced by war was not unique, that the demons they deal with have a long history' (Tritle 2004, 329).

Through his research into Clearchus, Tritle was also able to reflect on and contextualise his personal experiences of war. His article on Clearchus even begins: 'In the spring of 1970 I arrived in Vietnam as a young infantry lieutenant for a year's tour of duty...' (2004: 325). It can therefore be seen that the exploration of the theme of PTSD in ancient Greece allowed Tritle to adopt an autoethnographic approach and the resultant reflections on his personal experiences of Vietnam gives his academic scholarship a powerful personal dimension and depth of insight that non-veterans cannot replicate (see also Tritle 2000).

Perhaps, then, this is the real value of this and all previous studies of PTSD in ancient Greece. Not to demonstrate that taking a distanced existentialist stance can allow us to check off the latest diagnostic criteria against historical texts and thereby 'prove' that specific named individuals had PTSD and it must therefore be a universal part of the human psychological condition, but rather to recognise the universality of the shared humanity in the recognition of suffering that is displayed between Homer and Achilles, Xenophon and Clearchus, Shay and his patients, and even between Tritle, his students and himself. As Tritle succinctly puts it: 'Xenophon may not have understood exactly why Clearchus was troubled, but he clearly sensed that he was and that the reasons for this were to be found in his long years of service as a solider' (2004, 336).

Conclusions

Today PTSD is widely recognised by the medical

profession and society at large, including in popular culture. Although the term itself was only first coined only in relation to veterans of the Vietnam War, it had evidently been present during earlier conflicts. The medical terminology and clinical diagnosis of what is currently known as PTSD have changed over time and will undoubtedly change again in future. Given its shifting and elusive, yet pervasive, character in modern Western culture, finding evidence and formulating definitions for it in ancient Greece is inevitably going to be tricky.

It is neither possible nor appropriate to try and retrospectively 'diagnose' a historical or literary character with PTSD. There are numerous methodological reasons that make such pseudo-diagnoses impossible. Most important among these is the fact that socio-cultural factors affect how post-traumatic conditions are allowed expression in different societies and this introduces unknown variables across cultures and across time. We cannot easily predict or account for these when we seek to understand how post-traumatic psychological reaction formation may have presented itself in ancient Greece, although this question may warrant further research. Furthermore, the subjects of such discussions, whether genuine historical individuals or not, are ultimately just literary constructs onto which ancient authors have projected their personal and contemporary understandings of battle trauma and they therefore cannot be taken as objective medical evidence.

But this is precisely where the value of our texts lies, because if we approach our texts not with a list of medical criteria in hand, but with the general understanding that they are informed by an amalgam of the lived experiences of generations of individuals whose traumatic battle experiences went into forming the literary tropes and motifs used by ancient authors and based on their observations of the world around them, then we can read these works through that lens and reinterpret them accordingly. Such tropes and motifs drew their power from the fact that they were recognisable to the ancient audience that their authors were writing for and for whom battle traumatised individuals were not a rare occurrence but, on the contrary, were recognisably commonplace within the society around them. From this perspective it is possible to suggest that in ancient Greece PTSD, Conversion Disorder and other traumatic stress conditions were not rare but, in fact, completely ubiquitous.

Acknowledgements

I am very grateful to Dan Boatright and Stephen O'Brien for the invitation to speak at the Warfare and Society in the Ancient Eastern Mediterranean conference and to submit this paper to the edited proceedings. A version of this paper was first presented at the History of Medicine Day School at the Thackray Medical Museum in Leeds. The insightful and challenging questions of both audiences have helped me to refine the resulting paper and strengthen my arguments. I am especially grateful to my colleagues Phil Freeman, Fiona Hobden, Geoff Ibbotson, Christopher Tuplin and Susan Williams who so generously gave their time to discuss draft versions of this paper with me and make helpful comments and suggestions. Without their encouragement it would never have been finished. Thank you.

Bibliography

Barré, M. L. 2001. 'Wandering about' as a topos of depression in ancient Near Eastern literature and in the Bible. *Journal of Near Eastern Studies* 60.3, 177-187.

Christ, M. 2001. Conscription of Hoplites in Classical Athens. *Classical Quarterly* 51, 398-422.

Christ, M. 2004. Draft evasion on-stage and off-stage in Classical Athens. *Classical Quarterly* 54, 33-57.

Christ, M. 2006. *The Bad Citizen in Classical Athens*. Cambridge, Cambridge University Press.

De Blois, 2005. Plutarch's Lycurgus: A Platonic Biography. In K. Vössing (ed.), *Biographie und Prosopographie*. 91-102. Stuttgart, Franz Steiner.

DSM-IV-TR 2000. *Diagnostic and Statistical Manual of Mental Disorder Fourth Edition Text Revision*. Washington DC, American Psychiatric Association.

Erichsen, F. E. 1866. *On Railway and Other Injuries of the Nervous System*. London, Walton and Moberly.

Fischer, A. and Manstead, A. H. 1996. Emotions. In A. Kuper and J. Kuper (eds.), *The Social Science Encyclopaedia* 2nd ed., 239-240. London, Routledge.

Friedman, M. J. 1981. Post-Vietnam Syndrome. *Psychosomatics* 22, 931-942.

Gay, P. 1988. *Freud: A Life for Our Time*. London, Dent.

Greaves, A. M. 2009. Post-natal depression in ancient Greece. *Midwives* April/May, 40-41.

Greaves, A. M., 2010. *The Land of Ionia: Society and Economy in the Archaic Period*. Malden MA, Wiley-Blackwell.

Hanson, V. D. (ed.) 1991. *Hoplites: The Classical Greek Battle Experience*. London, Routledge.

Hanson, V. D. 2001. *Why the West has Won: Carnage and Culture from Salamis to Vietnam*. London, Faber.

Hinton, D. E., Pich, V. and Pollack, M. H. 2005. Panic attacks in traumatized Southeast Asian Refugees: Mechanisms and treatment implications. In A. M. Georgiopoulos and J. F. Rosenabum (eds), *Perspectives in Cross-Cultural Psychiatry*, 37-61. Philadelphia, Lippincott, Williams and Wilkins.

Holeva, V., Tarrier, N. and Wells, A. 2001. Prevalence and predictors of Acute Stress Disorder and PTSD following road traffic accidents: thought control strategies and social support. *Behavior Therapy* 32, 65-83.

Hornblower, S. 2000. Sticks, stones and Spartans: the sociology of Spartan violence. In H. van Wees (ed.), *War and Violence in Ancient Greece*, 57-82. London, Duckworth.

Johnson, H. and Thompson, A. 2008. The development and maintenance of Post-Traumatic Stress Disorder (PTSD) in civilian adult survivors of war trauma and torture: a review. *Clinical Psychology Review* 28.1, 36-47.

Kleber, R. J., Brom, D. and Defares, P. B. 1992. *Coping with trauma: Therapy, prevention and treatment*. Amsterdam, Swets and Zeitlinger.

Kruger, P. A. 2005. Depression in the Hebrew Bible: An update. *Journal of Near Eastern Studies* 64.3, 187-192.

Kulka, R. A., Schlenger, W. A., Fairbank, J. A., Hough, R. L., Jordan, B. K., Marmar, C. R. and Weiss, D. S. 1991. *Trauma and the Vietnam War Generation*. New York, Brunner/Mazel.

Labuc, S. 1991. Cultural and societal factors in military organizations. In R. Gal and D. Mangelsdorff (eds.), *Handbook of Military Psychology*, 471-489. New York, Wiley.

Lamprecht, F. and Sack, M. 2002. Posttraumatic Stress Disorder Revisited. *Psychosomatic Medicine* 64, 222-237.

Leese, P. 2002. *Shell Shock: Traumatic Neurosis and the British Soldiers of the First World War*. Basingstoke, Palgrave Macmillan.

MacDowell, D.M. 1995. *Aristophanes and Athens: An Introduction to the Plays*. Oxford, Oxford University Press.

McDonald, L. S., Bhasin, R. and Mollica, R. F. 2005. Project One Billion: A global model for the mental health recovery of postconflict societies. In A. M. Georgiopoulos and J. F. Rosenabum (eds.), *Perspectves in Cross-Cultural Psychiatry*, 303-29. Philadelphia, Lippincott Williams and Wilkins.

Morley, N. 1999. *Writing Ancient History*. London, Duckworth.

Morley, N. 2000. *Ancient History: Key Themes and Approaches*. London, Routledge.

Morley, N. 2004. *Theories, Models and Concepts in Ancient History*. London, Routledge.

Padel, R. 1995. *Whom Gods Destroy: Elements of Greek and Tragic Madness*. Princeton, Princeton University Press.

Raaflaub, K. A. 2007. Introduction: Searching for peace in the ancient world. In K. A. Raafalub (ed.), *War and Peace in the Ancient World*, 1-33. Oxford, Wiley/Blackwell.

Reber, A. S., and Reber, E. 2001. *The Penguin Dictionary of Psychology* 3rd ed. London, Penguin.

Richards, A. (ed.) 2004. *Report of the War Office Committee of Enquiry into 'Shell Shock'*. London, Imperial War Museum.

Rogers, B. B. 1910. *The Archarnians of Aristophanes*. London, George Bell and Sons.

Roth, S., Newman, E., Pelcovitz, D., van der Kolk, B. and Mandel. F. S. 1997. Complex PTSD in victims exposed to sexual and physical abuse: results from the DSM-IV Field Trial for Posttraumatic Stress Disorder. *Journal of Traumatic Stress* 10.4, 539-555.

Saul, L. S. 1945. Psychological factors in Combat Fatigue. *Psychosomatic Medicine* 7, 257-272.

Scott, L. 2005. *Historical Commentary on Herodotus, Book 6*. Leiden, Brill.

Shay, J. 1991. Learning about Combat Stress from Homer's *Iliad*. *Journal of Traumatic Stress* 4.4, 561-579.

Shay, J. 1994. *Achilles in Vietnam*. London, Scribner.

Shephard, B. 2000. *A War of Nerves: Soldiers and Psychiatrists*. London, Jonathan Cape Page.

de Silva, P. 1999. Cultural aspects of post-traumatic stress disorder. In W. Yule (ed.), *Post-Traumatic Stress Disorders: Concepts and Therapy*, 116-138. Chichester, Wiley.

Sommerstein, A. 1980. *Archarnian*. Warminster, Aris and Phillips.

Stirling, J. D. and Hellewell, J. S. E. 1999. *Psychopathology*. London, Routledge.

Tritle, L. 2000. *From Melos to My Lai: War and Survival*. London, Routledge.

Tritle, L. 2004. Xenophon's portrait of Clearchus: A study in Post-Traumatic Stress Disorder. In C. Tuplin (ed.), *Xenophon and His World: Papers from a conference held in Liverpool in July 1999*, Historia Einzelschriften 172, 325-339. Stuttgart, Franz Steiner.

Tritle, L. 2006. Warfare in Herodotus. In C. Dewald and J. Marincola (eds.), *The Cambridge Companion to*

Herodotus, 209-223. Cambridge, Cambridge University Press.

van der Kolk, B. A., Roth, S., Pelcovitz, D., Sunday, S. and Spinazzola, J. 2005. Disorders of Extreme Stress: The Empirical Foundation of a Complex Adaptation to Trauma. *Journal of Traumatic Stress* 18, 389-99.

van Wees, H. (ed.) 2000. *War and Violence in Ancient Greece.* London, Duckworth.

van Wees, H. 2004. *Greek Warfare: Myths and Realities.* London, Duckworth.

Weintraub, M. I. 1983. *Hysterical Conversion Reactions.* Jamaica NY, Spectrum.

www.ingramcontent.com/pod-product-compliance
Lightning Source LLC
Chambersburg PA
CBHW041709290426
44108CB00027B/2902